Introduction to Modeling in Wildlife and Resource Conservation

Introduction to Modeling in Wildlife and Resource Conservation

Norman Owen-Smith

Blackwell Publishing

BLACKWELL PUBLISHING
350 Main Street, Malden, MA 02148-5020, USA
9600 Garsington Road, Oxford OX4 2DQ, UK
550 Swanston Street, Carlton, Victoria 3053, Australia

The right of Norman Owen-Smith to be identified as the Author of this Work has been asserted in accordance with the UK Copyright, Designs, and Patents Act 1988.

1 2007

Library of Congress Cataloging-in-Publication Data

Owen-Smith, R. Norman
  Introduction to modeling in wildlife and resource conservation / by Norman Owen-Smith
    p. cm.
  Includes bibliographical references.
  ISBN 978-1-4051-4439-1 (pbk. : alk. paper) 1. Conservation biology–Computer simulation.
2. Wildlife management–Computer simulation. I. Title.

  QH75.O94 2007
  333.95′16–dc22
                              2006036893

A catalogue record for this title is available from the British Library.

Set in 10.5/12.5pt Photina
by Newgen Imaging Systems (P) Ltd, Chennai, India
Printed and bound in Singapore
by Fabulous Printers Pte Ltd

For further information on
Blackwell Publishing, visit our website:
www.blackwellpublishing.com

Contents

Preface

This book developed over more than a decade from a manual for a short course presented as part of a postgraduate program in conservation biology. Students arrived from various university backgrounds, generally with little or no prior experience in modeling, joined by some professional scientists seeking further training in the use of this information technology. The introductory course was aimed at providing those attending with the foundation from which to develop their own models, for applications that they might encounter later in their professional careers. The emphasis is not on mathematical formulae, but on the potential opened by computers and the software they may contain to explore, improve understanding, and contribute toward solving applied problems in wildlife conservation and resource management. My approach is to convey how dynamic equations generate patterns of change over time, subject to various influences, which can be viewed and comprehended in the graphical output. Through being inducted into the mind-expanding potential opened by modeling relationships, students should gain an appreciation of the heuristic value of modeling as an aid to learning, and hopefully also realize that developing a formula using a computer can be a fun activity.

My approach toward modeling has been strongly influenced by my first mentor, Tony Starfield; in particular his adage to keep models simple, if they are to serve effectively as a learning tool, perhaps discarding the model once it has achieved this aim. I was originally spurred to develop this teaching module after attending a short course presented by Michael Gilpin to induct South African conservation biologists into population models. He noted that conservationists need models more crucially than theoretical ecologists because they have to make decisions based on the best knowledge available to them at the time, without the opportunity for more research to resolve the uncertainties. The most defendable and transparent basis for such decisions is a formal synthesis of current knowledge and understanding, which is what a model represents. Through being exposed to criticism and subsequent improvement, the model contributes to improved, or at least shared, understanding.

The book benefited greatly from the critical feedback received from the many students and professionals who attended the biennial short courses over 14 years between 1992 and 2006. They cannot take responsibility for the additional chapters added, at the request of reviewers of the book manuscript.

Ward Cooper and three anonymous referees contributed greatly towards improving the original course manual submitted. Final production was helped greatly by Rosie Hayden and Sarah Edwards at Blackwell and Vijayan Anandan at Newgen Imaging Systems, while Phillip Tshabalala assisted with scanning some of the illustrations. I am grateful to John Lutz, owner of True BASIC, for permission to provide a limited version of this programming language on the CD accompanying this book. The True BASIC programming language is copyrighted by Lutz Services and no additional copies may be made without written permission of True BASIC. Additional information about other versions of True BASIC can be found at http://www.truebasic.com.

1

Introduction

Why learn modeling?

Chapter features

1.1   Introduction

All of us use *models* much of the time. Models are, by definition, simplified representations of reality. Hence we carry models of our world around in our heads and make decisions based on these models. The real world out there is somewhat more complex than our simplified perception of it, and hence our decisions may sometimes be wrong. Nevertheless, we can learn from mistakes and improve our conceptual model accordingly. Having gained experience, we may also want to communicate it to others. This entails translating the ideas in our heads into another medium – spoken words, text on paper like that you are reading now, a picture, a flow chart, numbers, or some other symbolic representation. In science, the formal language of mathematics provides a very powerful medium for communicating common understanding across linguistic barriers, for those inducted into it. For those less skilled in the symbolic logic of algebra and calculus, the desktop computer provides an invaluable aid. Indeed, modern computers extend the domain of mathematics, enabling solutions to be found for problems that had proved intractable to unaided logic. Tackling problems with the help of a computer can be a fun challenge and extend the capacity of our minds to think broadly. Computer models have become a crucial tool in addressing some of the big problems confronting humankind, like projecting future climates to be expected in a warming world, or how to

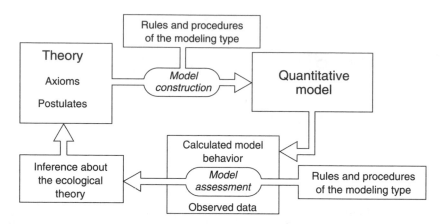

Figure 1.1   Formal representation of the construction of a model from ecological theory (from Ford 2000).

preserve a substantial proportion of the animal and plant species that are threatened by the expansion of our species across the globe. If you, as an aspiring ecologist, conservationist, or wildlife manager, want to make the best-informed and most defensible decisions about how to conserve or manage biological resources in the form of animal or plant populations, you should become competent in using a computer and the software it contains as an aid.

This book is intended to provide a starting foundation for this enterprise. By the end of it, you should feel empowered to develop the models and ideas it contains further, to address the real problems that you might encounter in your professional environment.

The *art* of modeling is to capture the system being modeled using just the most essential features. For example, a caricature is a pictorial model emphasizing the visual features that enable us to recognize a person. In *science*, models can be viewed as analogies, facilitating explanation by synthesizing knowledge and understanding (Ford 2000). They are conceptual constructs projecting the logical outcome of a set of starting assumptions and relationships (Fig. 1.1). The distinction between the real-world system being modeled and its formalized abstraction in the model world needs to be recognized (Fig. 1.2). Expressed as a set of functional equations, graphs, or box-and-arrow connections, a model becomes exposed to criticism, revision, and testing. It reveals the basic assumptions and logic that led to our expectations of certain outcomes. This conceptual model can then be elaborated into more formal mathematical or computational representation. Given the complexity of ecological systems, the connections between cause and effect are not always obvious. The model output can

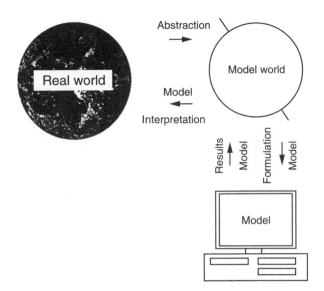

Figure 1.2   Relationships between the real world, the model world, and the computer representation of this model world (from Chapin et al. 2002).

often be surprising and contrary to initial expectations. This is where the heuristic or learning value of modeling arises: either a mistake was made in formulating the model, or there were connections that were not initially perceived. Modeling can resemble detective work, putting together the facts, clues, conjectures, principles, and other intellectual power at our disposal to understand and solve problems, as outlined by Hilborn and Mangel (1997).

The formal logic of mathematics has an important place in theoretical ecology, but modeling is more than mathematics. Analytic solutions often require assuming that conditions are spatially uniform and unchanging over time, so that equilibrium outcomes can be derived. But spatial heterogeneity and variability over time are pervasive features of ecological systems, which commonly exist far from equilibrium. Computational solutions can be obtained to problems that can't be solved analytically, albeit less elegantly. Hence, through mastering the power of your PC, you can explore beyond the scope of most mathematically oriented treatments of theoretical ecology. The same formulae placed in different environmental contexts can yield different outcomes, leading you into the "ecology of equations" (Owen-Smith 2002a).

Modeling is especially important in conservation biology, where we are concerned not merely with understanding, but rather with guiding decisions toward achieving certain ends or goals. Much is at stake, and there can be contrary opinions on the best course of action. The most defendable way of resolving such conflicts is via some transparent model that synthesizes current knowledge, and also accommodates uncertainty both in this understanding and in the conditions that might occur in the future. Models representing population dynamics have been most extensively developed. Animals and plants commonly provide resources for human benefit, including flesh, milk, or fruits for consumption, hides, timber, or medicine for other needs, forage to feed livestock, or less tangible goods, such as recreational hunting or wildlife viewing. Such bioresources are potentially renewable, provided the rate of extraction does not exceed the rate of regeneration. Nevertheless, the history of bioresource exploitation provides a dismal record of failure to sustain populations, for various reasons (Reynolds et al. 2001). An essential requirement for ecological sustainability is that underlying resources as well as the supporting environmental context remain intact and healthy. Hence to conserve large mammals or harvest fish, we need also to manage vegetation or maintain a productive marine environment. Accordingly, although the models developed in this book are focused especially on wildlife populations, represented by large mammals, birds, or fishes, models representing the dynamics of vegetation are also included.

The crucial first step in model development is to identify the purpose for which the model is being constructed (see Box 1.1 for a list). Several misconceptions often hold biologists back from embarking on modeling, as listed in Box 1.2. Make sure that you understand why these are misconceptions, by discussing them with your classmates or instructor, or by reading the article by Starfield (1997). A final common omission is neglecting to relate model findings back to the real-world problem for which the model was developed. The full iterative modeling procedure is outlined in Fig. 1.3, with several loops potentially traversed before an adequate solution is obtained. After modeling has been concluded, the final task is to

---

Box 1.1   Potential purposes of modeling

- Simplified description
- Synthesis of knowledge and understanding
- To guide experimental design
- To evaluate alternative scenarios
- As a basis for risk analysis
- As a tool for decision support
- As a basis for predicting the future.

Box 1.2   Misconceptions about modeling (from Starfield 1997)

1 A model can only be built once we have a complete understanding of the behavior of a system or population.
2 It is not useful to build a model if there are gaps in the data it is likely to need.
3 A model cannot be applied until it has been validated or proven to be accurate.
4 A model should be as realistic as possible, including the detailed intricacies of biological systems.
5 Modeling is an extension of mathematics, and so models cannot be understood by biologists unskilled in mathematics.
6 The primary purpose of modeling is to make predictions.
7 Modeling is a time-consuming and expensive activity.

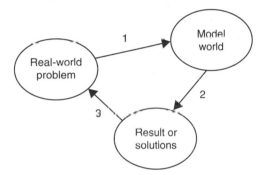

Phase 1   Create a model world containing a purposeful representation of the problem. Manipulate the model.

Phase 2   Report outcomes. Draw conclusions from the model world.

Phase 3   Relate the model results back to the real world. Reconsider whether the model formulation is satisfactory.

Figure 1.3   The iterative modeling process from the real-world problem to the model world and back via results or solutions to the real world again (from Beres et al. 2001).

communicate effectively what was learned, together with the assumptions and structural design that underpinned this enlightenment (Box 1.3).

1.2   Structure of the book

This is not a textbook on ecology, but rather a practical guide on how to formulate models for particular applications in wildlife conservation

---

Box 1.3 Protocol for disclosing the outcome of a modeling exercise (from Beres et al. 2001)

1 *Statement of purpose* Problem addressed. Intended users of the model.
2 *List of assumptions* The starting assumptions and their justification.
3 *Structural or design information* The type of model (demographic, statistical, geographic) and its format or "platform" (computer based, mathematical), including the simplifications made.
4 *List of variables, parameters, and relationships* Complete listing of these components, assumed forms of relationships, parameter values, units, degree of uncertainty, and the sources for these inputs.
5 *Decisions, features, uniqueness, and weaknesses* Deviations from standard models, alternative models rejected, acknowledgment of recognized weaknesses.
6 *Sensitivity analysis* Type of sensitivity analysis done and resultant ranking of parameters by their influence.
7 *Validation* Relationship between model output and real data.
8 *Scenarios* Combinations of parameters and inputs explored.
9 *Outcomes and interpretation* What was learned about the behavior of the "model world" and the conclusions drawn.
10 *Real-world consequences* Relating the findings from the model world back to the real-world problem, and an assessment of the reliability of the interpretation, suggestions for future directions in modeling or research.

---

and management. Only the basic theory needed for model formulation is covered in the chapters that follow. Some excellent books on ecological concepts are listed below. This book was developed from the manual for a 10-session course over two weeks, introducing beginning graduate students in conservation biology to modeling applications. The original 10 chapters have been expanded to include additional topics that some instructors might want to cover. Chapter 2 inducts students into thinking about modeling, by means of a hypothetical problem that encourages active engagement within groups. Other instructors might want to substitute a different case study. Chapters 3–5 cover the basic approaches to modeling population dynamics: descriptive equations, matrix algebra, and coupled consumer–resource interactions. Chapters 6–11 outline models applicable to particular topics in conservation or resource management. Chapters 12 and 13 illustrate modeling approaches to vegetation dynamics, considering plants both as a resource and as defining habitat conditions for animals. The final two chapters address two frontline topics: how to incorporate adaptive behavior into models and how to challenge models with data.

The models that will be developed in the course of the chapters remain fairly elementary. As emphasized by Starfield and Bleloch (1991), the main value of modeling is heuristic, that is, the knowledge and understanding gained in formulating the model, rather than the model itself. A special theme running through all of the chapters is how to accommodate temporal variability, which is a pervasive feature of real-world environments and not readily handled by mathematical abstraction. There is another book exploring the theme of environmental variability more comprehensively, with a specific focus on large mammalian herbivores and their interactions with vegetation (Owen-Smith 2002b). Textbooks going more deeply into aspects of modeling are listed below.

1.3 Supporting computer software

Models can be formulated in different software mediums. Spreadsheets are most easily accessible and have become familiar to most students by now. They are well suited to handling the numerical calculations involved in representing how populations change over time. Each chapter is supported by a corresponding appendix at the end of the book, guiding students into formulating models in Microsoft Excel. My approach is to start with a very elementary model, and then elaborate its concepts into more sophisticated versions. These models incorporate no frills, so that their workings remain transparent. It is important that students develop these models themselves, thereby gaining confidence in using the software, while learning the links from concepts and associated equations through the numbers produced to the patterns depicted in graphs.

Spreadsheets have limitations, particularly in representing the logical flow of more complex models in a transparent manner. Programming languages offer much greater flexibility in model development, and if written with appropriate structuring reveal more clearly the conceptual linkages. The programming language that I adopted is True BASIC (Kemeny and Kurtz 1990, copyrighted by Lutz Services), an advanced version of the original BASIC language modified to incorporate some of the new power opened by advances in computer technology. It offers several advantages as an introductory language for teaching purposes and can be written with literary clarity. Elementary programs can easily be produced with few formal constraints, and later be elaborated to take advantage of the structural clarity of subroutines and function definitions. At the same time, it offers powerful matrix operations. Students can comprehend the programs and thus modify them for specific applications, without needing more than an initial introduction to certain features of the language. True BASIC offers powerful graphics capabilities, but in practice it is generally more helpful to write the output graphed on the screen into structured text files for later retrieval into a spreadsheet, with its preprogrammed graphing capabilities. Most importantly for me, I can return to my programs after a year and still readily understand them.

Hence each chapter is supported additionally by models written as True BASIC programs. Some of these are direct counterparts of the spreadsheet models whereas others take advantage of the flexibility of the programming language to accommodate more elaborate features. These models can be found on the CD included with the book, together with some explanatory text files. Also included is a limited version of the True BASIC programming language, which can be used to modify existing programs or create new programs of 250 lines or less. Running more elaborate programs will require purchasing the full version of True BASIC language. Alternatively, a trial version can be downloaded from the Internet, which will function for a limited period (visit http://www.truebasic.com for more information about this and other options). Also included on the CD are versions of these same programs as stand-alone executable files, which can be run on any computer simply by double-clicking on them. The disadvantage of this route is that the programs cannot be altered, although many offer the opportunity to change parameter values interactively during the run. You can nevertheless inspect the text of the True BASIC programs that generated the output that you observe in the files with .tru extensions on the CD.

Alternative programming languages could be used, such as Visual Basic, R, PYTHON or C++, or modeling platforms, such as MATLAB, RAMAS, and STELLA. Each offers advantages and disadvantages in terms of ease of comprehension, pitfalls, and power. Instructors familiar with one of these alternative software packages will obviously have had some programming experience, and thus will be able to guide their students into translating the concepts incorporated in my True BASIC programs into their favored medium. The plan is to establish a website accessible via the Blackwell home page where versions of these same models written in another programming medium can be accessed – www.blackwellpublishing.com/owensmith.

Recommended supporting reading

The classical textbook guide to modeling applications in wildlife conservation is Starfield and Bleloch (1991). Burgman, Ferson, and Akcakaya (1993) provide a more general approach emphasizing some of the consequences of stochastic variability, whereas Williams, Nichols, and Conroy (2002) produced a massively comprehensive text encompassing both theoretical and statistical modeling of wildlife populations. Milner-Gulland and Mace (1998) summarize both ecological and economic theory underlying resource exploitation. Several textbooks provide helpful guides to the ecological theory incorporated into models, including Gotelli (1995), Hastings (1997), Case (2000) and Vandermeer and Goldberg (2003). Two general textbooks with good coverage of theory and its applications including supporting models are Begon, Townsend, and Harper (2006), and Sinclair, Fryxell, and Caughley (2006). Roughgarden (1998) illustrates concepts in theoretical ecology using models written in MATLAB software. Haefner (1996) provides a comprehensive outline of both mathematical and computational approaches to modeling across all fields of biology. Hilborn and Mangel (1997) present a philosophical guide to the statistical aspects of modeling, whereas McCallum (2000) outlines how to obtain and analyze the information needed for the construction of population models.

Supporting file on the CD

An introduction to programming in True BASIC.

2

A starting problem

Conservation of the dodo

Chapter features

Topics

Objectives; information; uncertainty; conceptual model; simplification; assumptions; risks

2.1   Introduction

In this chapter, you will be confronted with a conservation problem, at a stage when you have learned little about modeling except the underlying philosophy. With limited time available, it may be best to tackle it without the aid of a computer. The exercise is aimed at getting you to think about where modeling fits into problem solving at the start of the course. Your instructor might prefer to substitute some other problem for the one outlined below. You should attempt to work out your own answers (or those of your team) to the questions that will be posed, before turning to the guidance given in the appendix to this chapter at the end of the book.

The problem outlined below has been chosen because it raises some eternal issues in conservation biology, draws together several different strands to be considered in decision making, and offers some scope for imaginative solutions. It concerns a hypothetical rescue plan for an extinct bird called the dodo. Read on to obtain more information about what this involves.

2.2   Conservation of the dodo

2.2.1   Biology

The dodo (*Raphus cucullatis*) was a large flightless pigeon, which once inhabited the island of Mauritius in the Indian Ocean. As is well known, this bird

Figure 2.1   Drawing of a dodo, from Livezey (1993). Study of a dodo by F. Hart, 19th century. Royal Albert Memorial Museum, Exeter/Bridgeman Art Library, London.

became extinct early in the eighteenth century, and the cause was quite obvious. It provided food for early mariners on their way to the East Indies, and it was exterminated through human overkill plus the introduction of pigs and cats to its island habitat. You may not be aware that a similar bird called the solitaire (*Pezophaps solitaria*) once lived on the island of Rodrigues 600 km from Mauritius, and also became extinct through the agency of humans and introduced predators around the same time.

A feature article in the eminent science journal *Nature*, dated September 23, 1993, outlined attempts to "bring the dodo back to life" by reconstructing its morphology, behavior, and life history (Maddox 1993). The source of this review was a detailed paper in the *Journal of Zoology, London* by Bradley C. Livezey (1993) presenting "an ecomorphological review" of the dodo and solitaire (see Box 2.1).

2.2.2 Current conservation issues

Now let's turn to subsequent developments that will form the basis for the conservation problem that you must address. The Global Conservation Union (GCU) has just announced that an attempt to extract DNA from the organic remains of dodos preserved in museums has been successful. The complete genome of this extinct bird has been inserted into two turkey eggs, thereby initiating the development of two embryos. A conservation plan must urgently be formulated to restore a population of the species in

---

Box 2.1   Biology of the dodo *R. cucullatis* (from Livezey 1993)

Appearance   (extracted from a report by Strickland and Melville published in 1848)

> These birds were of large size and grotesque proportions, the wings too short and feeble for flight, the plumage loose and decomposed, and the general aspect suggestive of gigantic immaturity. We cannot form a better idea of it than by imagining a young duck or gosling enlarged to the dimensions of a swan.

Taxonomy   Member of Columbiformes, tentatively assigned to the family Raphidae together with the solitaire (*P. solitaria*).

Distribution   Restricted to the island of Mauritius in the Mascarene group of islands in the Indian Ocean, the dodo became extinct in the wild in the early eighteenth century and in captivity shortly thereafter.

Size   Estimated adult body mass – males 16–21 kg, females 13–17 kg. Large seasonal variations in fat deposition and hence mass are reported.

Habitat   Not stated.

Diet   Food stated to be "raw fruit," but probably also included seeds and leaves. The birds had a large crop with moderately large stones in the gizzard. Due to the current paucity of young plants of the waterberry trees (*Sideroxylon grandiflorum*), it has been suggested that this plant species was dependent on having its seeds pass through the digestive tracts of dodos to facilitate germination. However, this contention has been controversial.

Breeding behavior   Males and females formed persistent pairs (a "marriage") and actively defended territories extending 200 m from the nest with wing-rattling noises and attacks on intruders. The nest was a pyramid of palm leaves 45 cm high, and the male and female incubated in turn.

Life history   A single egg was laid, which was incubated for about 7 weeks before hatching. The young remained with its parents for "several months," during which time family groups from different territories mingled temporarily in aggregations of 30–40 birds. The parents thereafter remained as a "couple" for an unspecified period before re-nesting, whereas the young stayed together in groups. Longevity was estimated allometrically from body size to be 27 years.

---

the wild. The newly formed Dodo Rescue Group (DOREGRO) of the GCU has called for plans to be submitted on a competitive basis. The prize will be a generous 10-year grant to undertake the first field study on the biology of the dodo.

The DOREGRO has already purchased an uninhabited island near Mauritius to accommodate the anticipated dodo population. This island, called "Quattro," contains an abundance of giant waterberry trees (*S. grandiflorum*), upon the fruit of which the dodo was believed to depend. The area of this island is 151 ha. A nearby island called "Excel," covering 299 ha, could potentially also become available for stocking with dodos. However, there are people living on it, and they would have to be paid out for their land if they were to be moved elsewhere to make space for the dodo.

Everything that is known or inferred concerning the ecology and life history of the dodo is summarized in Box 2.1. That is all the biological information that you have at your disposal.

The specific questions that need to be addressed by the conservation plan are

1 How many dodos could be accommodated on the Quattro island?
2 How long will it be before the island is saturated with dodos?
3 If thereafter the population were to be harvested to supply zoos with specimens, so as to generate funds for the purchase of the Excel island, how many dodos per year could be sold?
4 If the alternative option of translocating dodos to Excel to establish a viable population there before selling birds to zoos was adopted, how long would it take before the purchase cost of the island of Excel could be recovered?
5 A third option would be to invest money in obtaining more genetic material to fertilize two additional eggs. This money would have to be recovered subsequently from the sale of dodos. How long would it take before the financial investment would be recovered?
6 A final important question is this: what is the risk that this whole endeavor will fail, that is, that the dodo population will die out? Consider the chances of this happening for the different options.

The financial information that you will need is given in Box 2.2. It is over to you to decide on the most appropriate population model to use. Proceed with this exercise as far as you can. If information is lacking, you will

---

Box 2.2   Estimates of financial costs involved

1 Cost of purchasing the Excel island – $10,000,000
2 Cost of repeating the genetic extraction, fertilization, and incubation – $50,000,000 per egg
3 Expected purchase price of dodos to be paid by zoos – $100,000 per bird

need to make assumptions. Tabulate all the assumptions that you made in developing your model.

After you have done your best to address the problem, turn to Appendices 2.1–2.3 for guidance about formulating an appropriate model, a list of some of the assumptions that need to be made, and what you should have learned from the exercise.

3

Descriptive models

Choosing an equation

Chapter features

Topics

Exponential growth; density dependence; continuous versus discrete time; difference versus differential equations; over- and under-compensation; time lags; thresholds; limit cycles; chaotic oscillations; Allee effect; environmental variability

3.1 Introduction

Changes in the abundance of an animal or plant population trace out patterns over time. Introduced into a new environment, a population grows rapidly at first, but later the rate of increase slows, and eventually some maximum abundance level is reached. Thereafter, the abundance may fluctuate somewhat from year to year, or even decline progressively, threatening eventual extinction.

Almost all of these patterns can be described by means of an appropriate equation. You have probably been introduced to the logistic equation. However, in scientific papers on population dynamics, you will find alternative equations invoked, labeled cryptically with peoples' names: Beverton–Holt, Ricker, Allee, Lotka–Volterra. What governs which equation, or modifications thereof, is most appropriate for describing the dynamics of the population that you might be studying.

The changes in the total population size (or numbers within some defined area, termed the population *density*) can be described directly as a function of time. However, it can be more revealing to consider how the *growth rate* of the population responds to changes in the population density. The growth rate can be described either as the numerical increment (or decrement) per unit time, or as a proportional (relative) rate of change. The latter reveals the *density dependence* of the growth rate, which is of fundamental importance in population ecology theory. Without density dependence, the population would either drift upward in numbers without limit, or downward toward eventual extirpation. In order for the population to persist, some process must counteract extreme trends, or *regulate* the abundance around some level. Regulation is synonymous with negative density dependence, that is, a reduction in the relative growth rate with increasing population density and vice versa. The shape of the relationship between the relative growth rate and the population density has an important influence on the trajectory followed by the abundance level over time.

A second crucial feature is how time is incremented. Dynamic processes are commonly formulated using the symbolism of differential calculus, that is, how the rate of change changes with infinitesimally small increments in time. However, populations are normally counted at fixed time intervals, for example, once annually. Moreover, the births and deaths governing changes in population abundance generally occur during restricted periods of the year. Accommodating this entails a switch from differential to *difference* equations using discrete time steps. This subtle change in how time is incremented can have a surprising effect on the dynamics generated, because a time lag is thereby built into the relationship between the changing population abundance and its effect on the population growth rate.

A third crucial feature, commonly neglected, is how the output dynamics respond to changes in the environmental conditions influencing population growth. Since variability over time is a key feature of real-world environments, it is important to understand how alternative equations accommodate the effects of such variability. Handling nonstationary conditions is a formidable task for a mathematician, but the consequences can easily be explored through numerical simulation using an appropriate equation.

This chapter will introduce you to a set of equations that has been used to describe population dynamics. The model is the equation, with its explicitly specified relationships. Explore the patterns that these equations generate using spreadsheet software, or turn to the prewritten True BASIC programs supplied on the accompanying CD. Doing both will enable you to compare the merits of these alternative computational mediums. This exercise will guide you into selecting the equation that most aptly represents the pattern of changes in abundance shown by any particular animal or plant population. Note that with all of these equations, the population size is described by one single number N, without taking into account the proportions in different age, size, or sex classes.

The sections below follow a learning progression, starting with some basic concepts expressed in the form of dynamic equations. From this foundation you will proceed from simple exponential growth through various ways of representing the density feedback restricting maximum population size. Finally you will consider how environmental variability affects the output generated by these equations.

3.2 Dynamic equations

An equation describes a relationship between two or more variables. The simplest is a straight line relationship between a dependent variable Y and some independent predictor X:

$$Y = a + bX,$$

where a sets the intercept with the Y-axis and b the slope as X increases. Curvilinear relationships can be depicted through a power function of X, for example,

$$Y = a + bX^2$$

with the value of Y curving upward as X increases in the form of a half-parabola (Fig. 3.1). However, this relationship can be made linear simply by taking logs on both sides

$$\log(Y) = \log(b) + 2\log(X).$$

Now the intercept is $\log(b)$ and the slope is 2. A curve rising asymptotically toward a maximum value, called a rectangular (or saturating) hyperbola,

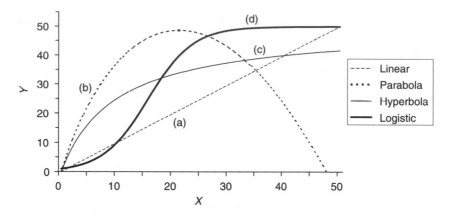

Figure 3.1   Shapes of the functional relationships described by particular equations: (a) linear $Y = a + bX$, (b) parabola $Y = aX - bX^c$, (c) saturating hyperbola $Y = aX/(b + X)$, and (d) sigmoid $Y = be^{aX}/(1 + ce^{aX})$.

is described by

$$Y = aX/(b + X),$$

where a defines the upper limit and b the value of X at which Y reaches half of its maximum. A sigmoid (S-shaped) curve is described by the logistic equation

$$Y = be^{aX}/(1 + ce^{aX}),$$

where e represents the base of natural logarithms, that is, the number 2.71828, and a, b, and c are the three parameters controlling the shape, upper limit, and initial elevation of the curve.

In population dynamics, interest lies in how the size of the population changes over time. Hence the dependent variable Y becomes the numerical size or density of the population usually labeled N, and X is replaced by t, representing time. Equations with time as the independent variable are called *dynamical* equations.

To describe changes in population size over very small time intervals, a formulation in differential calculus is appropriate, that is,

$$dN/dt = f(N, t),$$

where $f(N, t)$ is some unspecified function of the variables N and t. Note that we have switched to specifying how the growth rate changes as a function of the population size. The calculus notation is appropriate for the human population because births and deaths occur throughout the year. It is also suitable for populations of microbes growing in cultures where seasonal variation is irrelevant. The rate equation can then be integrated to obtain the total population size at any particular time, given its starting size.

However, for populations censused at discrete time intervals, commonly shortly after the birth pulse, a difference equation is more appropriate:

$$N_{t+1} - N_t = f(N, t),$$

where N_t represents the population size at time t. The discrete time formulation also corresponds with the way in which computers work (although continuous change can be approximated by making the time steps very short).

3.3   Geometric and exponential growth

The exponential growth model constitutes the starting foundation for population dynamics theory. It is a kind of null model, expressing what would happen if the population was free to grow at its intrinsic potential. This simple model was used by Reverend Thomas Malthus to project the consequences of unchecked human population increase back in the 1700s, thereby influencing Charles Darwin's formulation of the theory of evolution by natural selection. Hence exponential growth is sometimes termed Malthusian growth.

Let us start by deriving the exponential growth model from the simple accounting model that would have been appropriate for the dodo population considered in Chapter 2 (see Appendix 2.1). According to this formulation,

$$N_{t+1} = N_t + B - D,$$

where N_t represents the number (or numerical density) of organisms in the population at time t, B the births added, and D the deaths subtracted. For large populations, it is convenient to express B as a proportional birth rate b, where b equals B/N_t, and D as the corresponding relative death rate d, equal to D/N_t. Hence

$$N_{t+1} = N_t(1 + b - d).$$

The difference between the birth rate and the death rate is the effective proportional rate of population growth, which we can symbolize by r.

Hence we obtain

$$N_{t+1} = N_t(1 + r), \tag{3.1}$$

which means that the future population is equal to the current population plus some increment obtained by multiplying the current population by its proportional rate of increase.

We can take this equation a stage further by replacing $(1 + r)$ with the multiplicative growth rate λ (pronounced lambda), and rearrange the terms to get

$$N_{t+1}/N_t = \lambda.$$

Note that λ represents the factor by which the current population must be multiplied to get the future population one time step later. If r is positive, λ is greater than one and the population at the next time step will be larger than it is currently. If λ is less than one, the population declines. In some textbooks, you will find the symbol R used in place of λ. It actually symbolizes the population increase *per generation*, that is, there is a complete turnover of individuals between one time step and the next. This is appropriate for insects, plants, and other organisms with annual life histories.

The population size at any future time t can be obtained using the formula $N_t = N_0\lambda^t$, where N_0 represents the population size at time zero. This is geometric growth, that is, the population increases by a constant multiplier over a sequence of discrete time steps.

Equation (3.1) can be converted into a difference equation, by rewriting it as

$$N_{t+1} - N_t = rN_t,$$

or, in alternative terms,

$$\Delta N/\Delta t = rN_t.$$

where ΔN represents the change in population size over some discrete time interval Δt.

To represent true exponential growth, the time step is made infinitesimally small, yielding the calculus expression

$$dN/dt = rN, \tag{3.2}$$

where r , the "instantaneous" population growth rate, differs slightly from the r used for discrete growth above. You should recall from basic calculus that the future population size at any particular time t is obtained by integrating this equation, which yields $N_t = N_0 e^{rt}$. Exponential growth produces a somewhat larger population at a later time than geometric growth for the same value of r, just as compound interest gives a higher monetary return than simple interest. This is because offspring or interest earned contributes immediately to generating further growth.

If the relative growth rate r is quite small, say less than 0.05, then $(1 + r)$ is approximately equal to e^r, and the difference between geometric and exponential growth becomes very small. This provides a simple formula for calculating the doubling time for the human population. We want to obtain the value of t at which N_t becomes twice N_0. From the integral for exponential growth, we get $N_t/N_0 = e^{rt} = 2$. Taking natural logs, we obtain $rt = \log_e(2)$. Now the value of $\log_e (2)$ is 0.693 (check this answer in a spreadsheet). Hence to obtain the doubling time t, divide 0.693 by r. For example, at the end of the twentieth century the world human population was growing at the rate of 1.4% per year, so that $r = 0.014$ per year. This projects a doubling time of 50 years. However, because the growth rate is actually slowing, doubling is likely to take longer.

Note the important distinction between "projection" and "prediction." The projection forecasts what the outcome would be if nothing changed, in particular, if parameter values remained constant. For prediction, you would need to take into account all the things that might happen. This is why weather forecasters do not claim to predict tomorrow's rainfall, but instead merely provide a forecast of the probability that rain will fall.

Since r is expressed as a proportion per unit time, when the time dimension is changed, the value of r must be altered accordingly. For example, the instantaneous growth rate per week in the exponential growth model is simply the annual growth rate divided by 52, that is, the number of weeks in a year. For geometric growth, the weekly growth multiplier is $\sqrt{1/52}$ of the annual multiplier, that is, $\lambda^{0.0192}$, where $0.0192 = 1/52$.

Now is the stage at which to stop confronting numerical theory and turn to the aid of a computer to explore further. Appendix 3.1 provides guidance on how to formulate the geometric growth equation in a spreadsheet and graph the output. Plotting changes in population size against time, you should observe an upwardly rising curve, indicating that the number of individuals added to the population each year gets increasingly greater (Fig. 3.2). Are you surprised at how large the population has become after 50 years? Transform the Y-axis to a log scale, and you should see a straight line, indicating that the *proportional* rate of increase remains constant. The slope of the line yields the instantaneous growth rate r (equal to $\log_e(\lambda)$). Alternatively, plot the proportional rate of change, that is, the annual increase in the population size divided by the population size, against the increasing total population N. This should

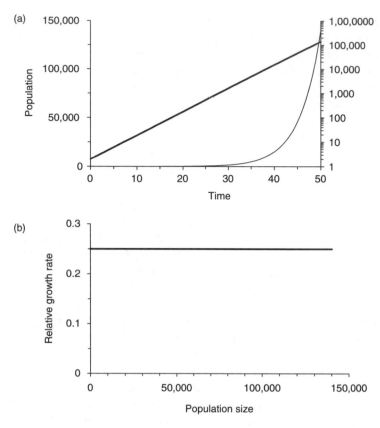

Figure 3.2   Geometric (discrete time exponential) growth model showing (a) change in population size over time, on both the arithmetic and the log scale and (b) relative population growth rate against population size.

yield a straight line parallel to the X-axis, intersecting the Y-axis at the value of r.

3.4   Adding a population ceiling

Exponential growth cannot continue indefinitely because ultimately space will become limiting if nothing else does. In the dodo example, territory size placed an upper limit on the number of breeding pairs that could be accommodated, although not necessarily on the total population size if nonbreeding birds could move through these territories. These "floaters" fill vacancies arising when breeding birds die although many will perish through harassment or lack of resources. Nevertheless, the proportion of the population constituted by floaters will ultimately be limited by recruitment into this segment from the offspring produced by breeding pairs.

The ceiling model recognizes the existence of some upper limit to population size in a very simple way. The population continues growing exponentially until its abundance reaches this limit, at which point further grow halts abruptly. This entails incorporating a condition in the equation:

$$N_{t+1} = \min(\lambda N_t, K),$$

meaning that the value of N_{t+1} is either the product of the population growth factor and the population size, or the maximum abundance K, whichever is smaller. Appendix 3.3 explains how this condition is formulated using a logical (IF ...THEN ...ELSE) function in a spreadsheet. Figure 3.3 shows the output that you should obtain. The ceiling model

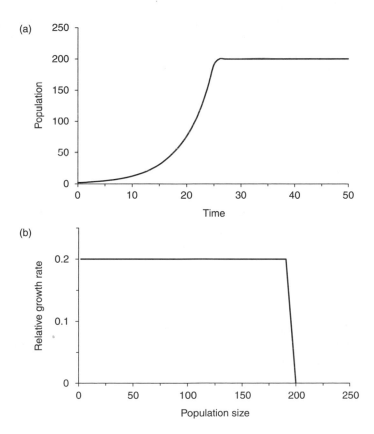

Figure 3.3   Ceiling model showing (a) change in population size over time and (b) relative population growth rate against population size.

is appropriate when no information is available on how the population growth rate changes with increasing density below the maximum size.

Although this model appears rather crude, it is widely used, especially in situations where the population is harvested to keep it below the upper level that might otherwise be reached. All that one needs to know then is the approximate growth rate r for the range in density levels that is maintained. Nevertheless, it is helpful to specify the ceiling to ensure that the modeled population does not become unrealistically large if harvesting is suspended or unusually favorable conditions cause the population to increase substantially despite the offtake.

3.5   Basic density-dependent models

More realistically, the population growth rate slows gradually rather than abruptly as the population size approaches the upper limit to the density that the area can support. Resource limitations, or other influences, progressively reduce birth and survival rates, until eventually the death rate matches the birth rate and further population increase ceases. As a result, we expect the population abundance to follow an S-shaped or *sigmoid* trajectory over time (Fig. 3.4a). Various equations may be used to describe such a pattern.

3.5.1   Logistic model

The most widely used equation generating a sigmoid pattern of population growth is the logistic equation. This is generally written in calculus notation as a differential equation describing how the change in population abundance or density over time depends upon the population size:

$$dN/dt = r_0 N (1 - N/K), \tag{3.3}$$

where r_0 represents the maximum proportional growth rate when the population size is very small ($N \sim 0$), and K represents the maximum population size attained, generally termed the "carrying capacity." Multiplying through, this equation becomes

$$dN/dt = r_0 N - (r_0/K)N^2. \tag{3.4}$$

Due to the squared term, the numerical population growth increment is a parabolic function of N. A plot of the change in population size with time, dN/dt, versus N yields a curve with a peak when $N = 1/2K$ (Fig. 3.4b).

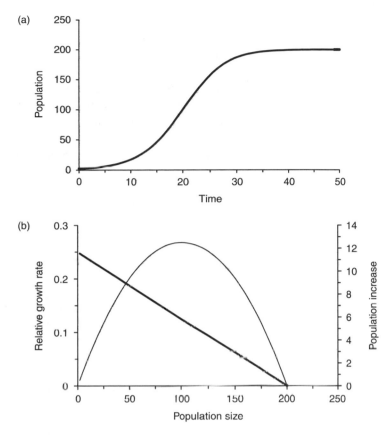

Figure 3.4 Logistic model showing (a) change in population size over time and (b) relative population growth rate as well as the annual population increment against population size.

Integrating eqn (3.3) gives the following expression for how the total population size changes over time:

$$N = K/(1 + e^{a - r_0 t}), \tag{3.5}$$

where a is the constant of integration defining the position of the curve relative to the origin. Plotting this function generates a smooth sigmoidal curve, with K setting the ceiling to the population size, that is, the asymptotic abundance level at which the population growth rate becomes zero (Fig. 3.4a).

However, most revealing is a plot of the proportional (or relative) growth rate in relation to the population size N. Dividing both sides of eqn (3.3) by N, we get

$$dN/Ndt = r_0(1 - N/K), \tag{3.6}$$

which is the equation for a straight line with Y-intercept given by r_0 and X-axis intersected at K (Fig. 3.4b). The proportional (or relative) growth rate declines linearly with increasing population size with a slope of $-r_0/K$.

For computation, eqn (3.3) needs to be transformed into discrete time units, yielding

$$\Delta N/\Delta t = r_0 N(1 - N/K).$$

Since $\Delta N/\Delta t = N_{t+1} - N_t$, we get, after rearranging

$$N_{t+1} = N_t(1 + r_0(1 - N_t/K)). \tag{3.7}$$

This equation is known as the "logistic map," that is, it maps or transforms the earlier population size into the population size one time step later, following the functional form of the logistic equation.

Set up the discrete time logistic model in a spreadsheet, referring to Appendix 3.3 for guidance. Explore the consequences of choosing alternative values for the parameters r and K. Note the following features in the graphical or numerical output:

1 Provided the value for the maximum growth rate r_0 is not too large ($r_0 < 1$, meaning the population size less than doubles in any time step), the population tends smoothly toward the maximum density K.
2 The annual increase peaks when population size is equal to $1/2K$.
3 The relative growth rate declines linearly to reach zero when N equals K.
4 The initial population growth rate is negative if the starting population size exceeds K.

Despite its simplicity, the discrete time logistic equation can generate surprisingly diverse dynamics. Increase r_0 to a value between 1 and 2, and note how the population size fluctuates before settling at its equilibrium abundance K. Raise r_0 further to a value between 2 and 2.5. Now persistent oscillations develop with a constant amplitude and period, called the stable limit cycle. The population overshoots carrying capacity, falls back below K, then rises above K again, *ad infinitum*. If r_0 exceeds 2.7, the population size jumps about in a seemingly erratic or *chaotic* pattern (Fig. 3.5). If r_0 is

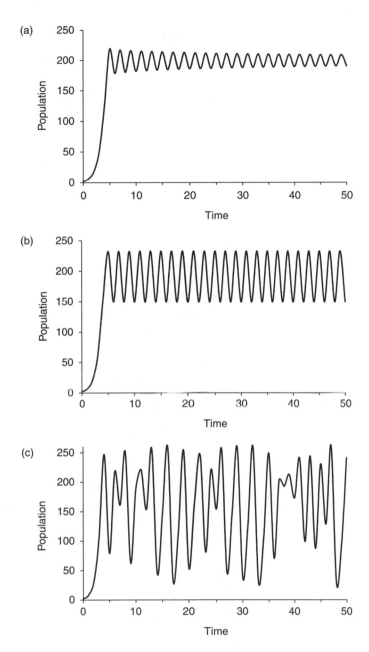

Figure 3.5 Output generated by the discrete time logistic model for high values of the intrinsic growth rate r_0: (a) dampened oscillations, $r_0 = 2$; (b) sustained limit cycle, $r_0 = 2.2$; (c) chaotic fluctuations, $r_0 = 2.9$.

made larger still, the computed population size may descend into negative numbers, then return from this imaginary realm, in a quite unrealistic way. Furthermore, small changes in the initial population size generate different output patterns. Nevertheless, the plot of the relative growth rate against the population size still remains effectively linear.

The seemingly chaotic behavior that can be generated by nonlinear difference equations has fascinated mathematicians. Gleick (1987) provides an interesting exposition of the discovery of chaos by mathematical modelers working in different scientific disciplines, from ecology to meteorology, and May (1981) gives a good overview of the chaotic patterns generated by the discrete time logistic equation.

The apparently chaotic output arises as follows. A difference equation incorporates an automatic time lag between cause and effect because the population increase is calculated from the population size one time step earlier. If the population growth rate is sufficiently high, the population can jump from well below K to well above K in one time step. The density effect then makes the growth rate negative, so that the population drops back below K again. With steep enough density dependence, small differences in population size have greatly amplified effects on the annual growth increment.

You can see the effect of the time lag simply by reducing the time step in the model, say from annual to monthly. Remember to reduce the maximum growth rate r_0 according to the changed time unit. The regular or chaotic oscillations disappear because the feedback is now generated by the population size one month earlier rather than a full year earlier.

3.5.2   Ricker or exponential logistic model

Since the logistic map incorporates a subtraction, N_{t+1} can become less than zero if N_t exceeds K by a sufficient amount, which is likely to happen when the potential population growth rate r_0 exceeds 3 per unit time step. This inconvenient feature might be of little concern for populations with slow annual rates of increase, like most large mammals, but is problematic for organisms that can potentially grow much more rapidly, like many fish. Hence fisheries biologists have favored an alternative difference equation called the Ricker model, which has also come into widespread use for large mammals (e.g., by Taper and Gogan 2002).

This equation differs from the logistic map through incorporating the density feedback within an exponential coefficient:

$$N_{t+1} = N_t \exp[r(1 - N_t/K)], \tag{3.8}$$

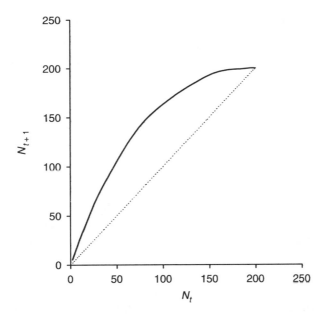

Figure 3.6   Ricker model output – plotted as a map of N_{t+1} against N_t.

where exp represents e, the base of natural logarithms, the term in brackets is the power to which e is raised, and $r = \log_e (1 + r_0)$. Note the similarity with the exponential growth equation. Rearranging and then taking natural logs on both sides of the equation, we get

$$\log_e(N_{t+1}/N_t) = r(1 - N_t/K).$$

Hence the relative population growth rate, calculated now as a log-transformation of the population change expressed as a ratio, decreases linearly with increasing population density. The important difference is that, because we are now multiplying the current population size N_t by an exponential term, the population size can never become negative, only infinitesimally small, because e raised to a negative number is still positive. This model is supposedly appropriate for fish populations where density dependence arises directly through cannibalism from adults on eggs and larvae. Its output closely matches that from the discrete time logistic. Confirm this on a new worksheet in your expanding folder of spreadsheet models. Textbooks also commonly show the output of the Ricker model in the form of a map of N_{t+1} versus N_t (Fig. 3.6).

3.5.3 Beverton–Holt or hyperbolic growth model

Another model frequently used by fisheries biologists, which likewise cannot generate negative numbers, is the Beverton–Holt equation:

$$N_{t+1} = \lambda N_t/(1 + aN_t). \tag{3.9}$$

In this model, the effective value of the finite population growth factor λ gets reduced as the value of N_t in the denominator rises. The zero growth level is set jointly by the values of the parameters λ and a; specifically $K = (\lambda-1)/a$. Hence raising λ also elevates the effective carrying capacity. This is not unrealistic if, for example, the growth rate is increased through a reduction in density independent mortality caused by a predator. This model is supposedly a better representation of fish populations where recruitment limitation arises mainly through competition among the larvae. It has also been widely used to model insect populations. A claimed advantage is it does not assign a special meaning to a "carrying capacity." This is merely the abundance level at which the relative growth rate reaches zero (i.e., λ becomes 1.0), as determined by the joint influence of the parameters a and λ on the slope of the density effect.

Set up the hyperbolic equation in a new worksheet. You may find that being unable to specify the maximum population size is somewhat inconvenient. It is not intuitively obvious that a needs to be made very small (e.g., 0.01 or less), otherwise the population can't grow much at all. To enable the settings of r and K to be specified as was the case for the logistic model, make $\lambda = 1 + r$ and $a = r/K$, in eqn (3.9).

The output from the Beverton–Holt model appears closely similar to that from the logistic and Ricker models, except that a slight concavity is evident when plotting the relative growth rate as a function of the population size (see Fig. 3.7). This small difference has surprising consequences. The hyperbolic equation will not generate oscillatory dynamics, even for high values of r. Raising the value of the maximum population growth rate merely increases the concavity in the response of the growth rate to increasing density. The resultant gradual approach toward the zero growth level avoids overshoot of the maximum density, however fast the population grows initially. This peculiar feature can be suppressed by a small modification, as you will see in the next section.

3.6 Curvilinear density dependence

In reality, the relationship between the relative growth rate and population size need not be linear. Rising population density may at first have little

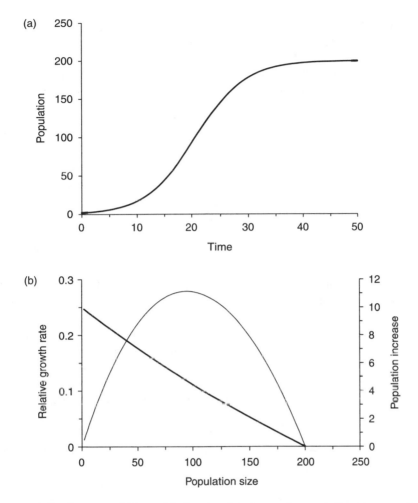

Figure 3.7 Beverton–Holt model showing (a) change in population size over time and (b) relative population growth rate plus annual population increment against population size.

influence on the population growth rate because food remains plentifully available. Toward higher density levels, resource competition may become effective, causing a fairly abrupt decline in the growth rate toward the carrying capacity level. This convex (when viewed from above) form of density dependence is believed to be typical of large mammal populations, both marine and terrestrial (Fowler 1987). In contrast, a concave pattern whereby the population growth rate declines sharply at first, but more gradually toward higher population levels, is believed to reflect a population limited by predators, which become saturated with more prey than they can consume at high prey abundance. The latter pattern seems to be a common feature for insect populations.

The Beverton–Holt model can be generalized to describe any form of curvilinearity in the density dependence by adding a power coefficient to the denominator:

$$N_{t+1} = N_t \lambda / (1 + [aN_t]^b). \tag{3.10}$$

Convex density dependence is generated by setting $b > 1$, whereas concavity prevails if $b \leq 1$. The value of b also affects the population level at which the population increase rate peaks. Through making b sufficiently large, relative to the value of λ, oscillations, cycles, and even chaos can now be generated. Explore this by incorporating b into the equation for the Beverton–Holt model in your spreadsheet, thereby transforming it into a "generalized sigmoid" equation.

The useful feature of this expanded model is that the value of b can be adjusted to represent any observed degree of curvature in the density relationship (Fig. 3.8). Its value determines whether the density feedback is "overcompensating" or "undercompensating" in the vicinity of the zero growth level. Overcompensation occurs when a small overshoot of carrying capacity brings about a greater reduction in the population size below the carrying capacity level in the next time step. This generates oscillations in abundance around the carrying capacity level. With the logistic model, you found that unstable dynamics (cyclic or chaotic fluctuations) resulted when the maximum population growth rate was sufficiently high, which automatically caused the slope of linear density dependence to be steep. Now you have discovered that population instability can also arise when curvilinear density dependence generates a sufficiently sharp decline in the population growth rate in the zero growth region, even if the maximum population growth rate is not that high. Explore this further by adjusting the values for λ (i.e., $1 + r$) and the power coefficient b in the generalized Beverton–Holt model. On the other hand, models with b less than one do not generate population oscillations, even when the maximum population growth rate is high, due to their gradual "undercompensatory" approach to equilibrium.

Nonlinear density dependence can also be produced by adding a power coefficient to the logistic equation, following Gilpin and Ayala (1973):

$$dN/dt = rN\{1 - (N/K)^\theta\}. \tag{3.11}$$

Boyce (1989) used this "theta-logistic" equation in formulating management models for elk in North America. However, like the logistic map, it

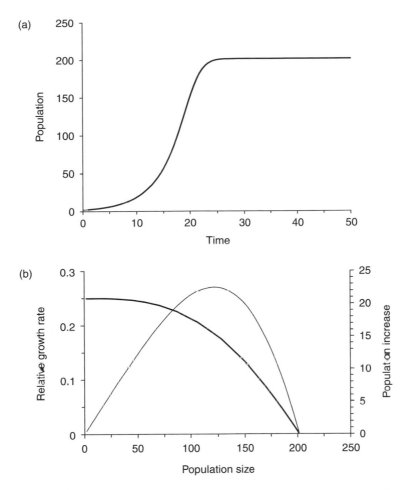

Figure 3.8   Generalized sigmoid model output with power coefficient $b = 3$ showing (a) change in population size over time and (b) relative population growth rate plus annual population increase against population size.

has the drawback in its discrete time version that negative numbers for N can be generated if the value of θ is sufficiently large, even when the population growth rate r is not that great. Modify the logistic model in your spreadsheet to incorporate the θ coefficient, and explore the consequences of different combinations of r and θ.

As an alternative to curvilinear density dependence, a threshold population level above which density dependence starts taking effect can be specified, while retaining a linear decline in the population growth rate beyond this level. Just two small modifications of the logistic equation are

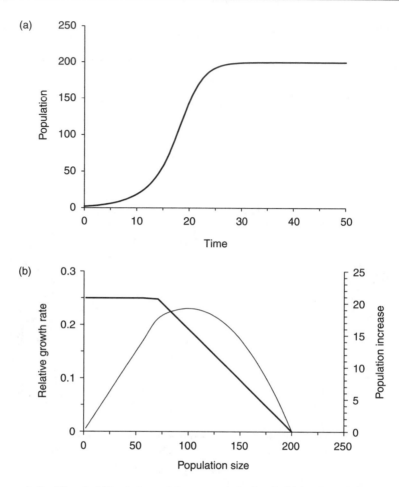

Figure 3.9   Threshold logistic model output with threshold level set at 70 showing
(a) change in population size over time and (b) relative population growth rate plus
annual population increment against population size.

needed: replace N by $(N - A)$, and K by $(K - A)$, where A is the threshold
density level (Fig. 3.9). Threshold density dependence gives a higher maximum increase rate than obtained from the simple logistic model for the
same value of r. Since the density feedback no longer accelerates with rising density as is the case for convexly curvilinear models, the threshold
model is less prone to generate oscillations, unless the threshold level A
is set very close to K. This model seemed to give the best approximation
to the form of density dependence displayed by a kudu population, as an
outcome of how different demographic segments were affected by rising
density (Owen-Smith 2000).

3.7 Delayed density dependence

The discrete time models that you have been formulating automatically relate population change to the density level one time step earlier, which with an annual iteration implies the previous year. However, if the density feedback arises through an interaction with food resources or predators, the effect of density could be delayed for longer than a year. For example, a high density of a prey species this year could result in an increase in the predator population the following year, and a consequent impact on the prey population the year after that. This assumes that populations largely turn over from one year to the next; in other words that the generation time is around one year, which is the case for most insects and many small mammals.

For large long-lived mammals, the lag in the density feedback could be somewhat longer than a year because of the extended generation times of such species. For example, after a high density of ungulates has degraded the vegetation, recovery could take several years. Furthermore, animals born under conditions of high density could grow up stunted, and hence less resistant to predation, resource limitations, and weather extremes, producing "cohort effects" (Prout and McChesney 1985). Mothers that are nutritionally deprived could also have weakened progeny resulting in "maternal effects," affecting not only large mammals (Boonstra and Boag 1987; Ginzburg and Tanneyhill 1994). As a consequence, the population growth rate could be influenced by the population density level several years back in the past. The model used by the International Whaling Commission for whale population dynamics assumes only the birth rate to be density dependent, but with a time lag equal to the period between birth and reproductive maturity, which ranges from 6 years for small Minke whales to 25 years for sperm whales (May 1981).

Delayed density dependence automatically leads to an overshoot of carrying capacity and consequent population oscillations (Fig. 3.10). If the lag between cause and effect is long enough, even populations with a low intrinsic growth rate show cyclic or chaotic oscillations. Explore this by introducing an explicit time lag into the logistic or any other density-dependent equation, through making population growth a function of population density two or more time steps earlier. May (1981) analyzed how the effective lag time relates to the natural response time of the population, equal to $1/r_0$. When the product of the time lag and r_0 is greater than e^{-1}, damped oscillations develop. If this product is greater than $\pi/2$, stable limit cycles occur, with a period roughly equal to four times the lag time. Test these predictions in your spreadsheet model. Bryant et al. (1991) suggested that the 10-year cycle of snowshoe hares and lynxes in northern

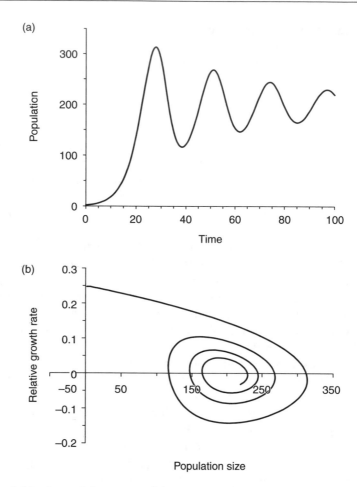

Figure 3.10 Lagged logistic model output with lag set at 4 years showing (a) change in population size over time and (b) relative population growth rate against population size.

Canada could be due to the 2–3 year delay in the recovery of birch shrubs from past browsing pressure by the hares. However, other evidence suggests that the predator–prey interaction with the lynx is primarily responsible for the cyclicity (Krebs et al. 1995).

3.8   Depensation or Allee effect

Depensation arises when the relative population growth rate declines toward low density levels. This is called the "Allee effect," after the ecologist who first recognized it. The mechanism could be difficulty in finding mates, lessened security against predation when groups become small or loss of the social facilitation that some species need for reproductive success. An example of the last is the passenger pigeon, which thronged the skies of North America in vast numbers, then suddenly collapsed to extinction. Hunting pressure alone seemed inadequate to explain the demise of such an abundant species. The Allee effect has important implications for conservation because it means that extinction could be predisposed once a population has dropped below some low abundance level, well before the last few individuals remain. Sufficiently strong depensation could even cause population growth to become negative at low density, leading inevitably to extinction. See Stephens and Sutherland (1999), and Courchamp, Clutton-Brock, and Grenfell (1999), for recent reviews.

Various equations could be used to generate depensation at low density through subtracting a suitable power term from logistic growth. One possibility is to use a quadratic polynomial, following Lewis and Kareiva (1993):

$$N_{t+1} = N_t[1 + r(1 - N_t/K)\{(N_t - b)/K\}], \qquad (3.12)$$

where r sets how fast the population will grow at its maximum rate, b sets the population level below which population growth becomes negative, and K sets the carrying capacity. An alternative equation with less extreme depensation was suggested by Edelstein-Kechet (1988):

$$N_{t+1} = N_t[1 + r\{(1 - a) - (1 - 3a)N_t/K - 2a(N_t/K)^2\}]. \qquad (3.13)$$

In this equation the strength of the Allee effect is controlled by varying the value of a between 0 and 1 (Fig. 3.11).

3.9   Incorporating environmental variability

For all of the models considered thus far, the output represents idealized circumstances in which environmental conditions remain constant. In the real world, the weather, food resources, predation risk, and other factors vary quite widely between years. For organisms like fish and insects with high reproductive potential, large numbers of offspring may be successfully recruited in some years, whereas very few survive in other years, depending on temperature and humidity conditions around the time of hatching, together with the abundance of predators and parasites feeding on these recruits. For African savanna ungulates, good rains can produce

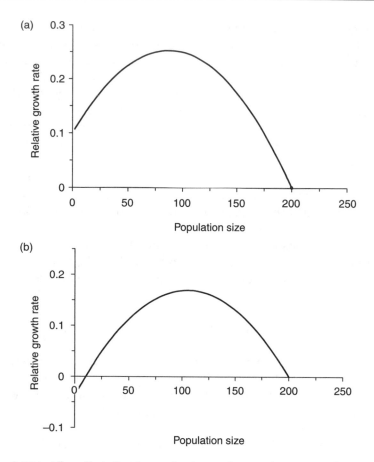

Figure 3.11   Allee effect, that is, a reduction in the population growth rate toward low density levels modeled using (a) eqn (3.13) and (b) eqn (3.12).

a vast superabundance of forage in some years, but insufficient food to last through the dry season in drought years when the rains fail. For northern ungulates, deep snow can restrict access to herbage during winter, also leading to widespread starvation during this critical period. There may even be cyclic patterns in these weather features, with periods when several benign years occur in succession followed by times when droughts or deep snows recur in several successive years.

Cyclic variability in the capacity of the environment to support the population, underlaid by fluctuations in food production or availability, can be modeled by making K a sine wave function of time:

$$K' = K\{1 + 0.5 \, \text{cosine}(2\Pi t/\tau)\},$$

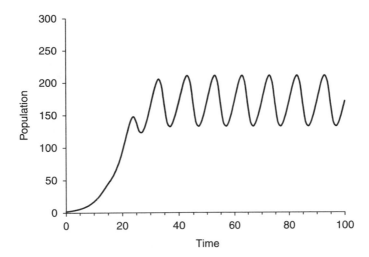

Figure 3.12   Environmentally entrained population oscillations driven by sinusoidal variation in the carrying capacity K with a period of 10 years.

where Π (Pi) $= 3.1416$ and τ is the time period between successive peaks in the habitat conditions. This equation makes the effective carrying capacity K' fluctuate from 50% above to 50% of the mean carrying capacity K, with a period between peaks in population abundance likewise equal to τ (Fig. 3.12). This represents environmentally entrained oscillations.

For many situations, the fluctuations in weather from year to year, especially in rainfall, seem to be essentially random. Both spreadsheets and programming languages have a function enabling a pseudorandom number ranging between zero and just under one to be obtained. For guidance on how to make carrying capacity K a random variable, refer to Appendix 3.4. The modeled variation in rainfall is somewhat unrealistic in that extremely high and low values are just as likely as middling ones (Fig. 3.13). The alternative would be to use some probability distribution for the environmental variations or actual measurements of the environmental factor believed to drive the population changes.

Explore the consequences of annual variability in the effective carrying capacity for each of the density dependent models, especially those introducing convexity in the density relationship. What patterns do you observe, and how realistic do they seem? For what combinations of r and θ or b can the population persist, that is, not descend into negative numbers? Examine in particular the graphs showing the relationship between the relative growth rate and population size. What are the implications for trying to detect density dependence from real-world data (see, e.g., Hassell, Latto,

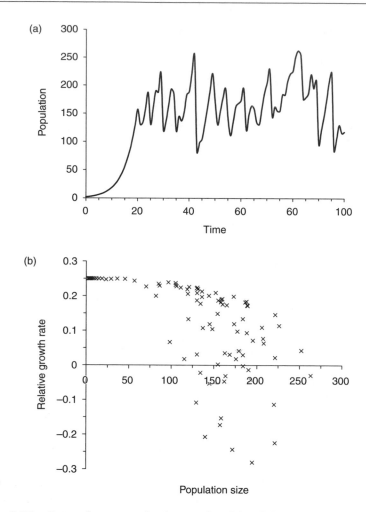

Figure 3.13   Output from generalized sigmoid model with $b=3$ and random variation in carrying capacity K between 0.5 and 1.5 of the mean K, showing (a) change in population size over time and (b) relative population growth rate against population size.

and May 1989, and Woiwod and Hanski 1992). Calculate the mean population density over some extended period after the peak abundance was reached, and note that this mean is less than the mean value of K that was set at the top of the worksheet. Try to work out why this is so.

3.10   Overview

All of the models that you considered in this chapter were simple two or three parameter equations. One parameter governed the maximum population growth rate. Another set either the zero growth level (e.g., logistic equation) or the slope of the decline in the growth rate with increasing density (e.g., Beverton–Holt equation). These amount to the same thing because the density feedback determines the population size at which the growth rate becomes zero. A third parameter controlled either the form of the density dependence (concave or convex) or some other feature (e.g., time lag). The main feature distinguishing these equations was how the density feedback on population growth was manifested. You also explored how the density feedback interacted with environmental variability to generate different outputs from equations that might project closely similar dynamics in a constant environment.

You now have the tools to formally describe almost any observed pattern of population dynamics (see Fig. 3.14, noting that the form of the trend line fitted depends on whether the data extend into the negative population growth region). You should have learned the following from the exercise:

1 Descriptive models can be expressed as simple mathematical equations.
2 Models differ mainly in the form of density dependence that they describe.
3 Almost any pattern of population dynamics can be represented by means of an appropriate equation.
4 Lagged effects inherent in the discrete time formulation can make a big difference to the model dynamics.
5 Models can incorporate environmental variability through its effects on either the ceiling density level or the maximum growth rate.
6 Models that might appear very similar in a constant environment can generate rather different dynamics when wide environmental variability is allowed.
7 Populations showing strongly overcompensating density dependence may not persist in highly variable environments if the inherent population growth rate is too high.

Recommended supporting reading

Gotelli (1995) and Case (2000) provide helpful outlines of some of the theoretical concepts such as density feedbacks incorporated into population models. Milner-Gulland and Mace (1998) explain some of the theoretical principles underlying the sustainable use of biological resources.

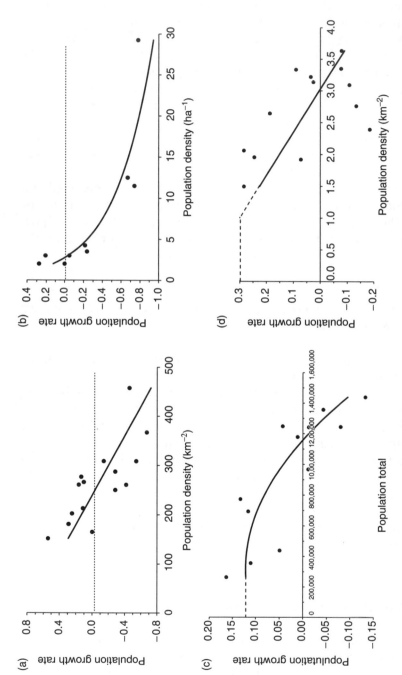

Figure 3.14 Some examples of density dependent growth rates exhibited by particular animal populations. (a) Linear density dependence, magpie goose (redrawn from Bayliss 1989), (b) concave density dependence, arctic ground squirrel (redrawn from Karels and Boonstra 2000), (c) convex density dependence, wildebeest (data from Mduma et al. 1999), and (d) threshold linear density dependence, kudu (from Owen-Smith 2000).

Programs on the accompanying CD

EXPONEN; LOGISTIC; DESCRPOP

Exercises

1 Modify one of the models so as to make it applicable to an animal or plant species of special interest to you, by setting appropriate functional forms and parameter values. Justify your choice of parameters and suggest possible uses of the model.
2 Explore published papers to find out whether the suite of models covered in this chapter is adequate to describe all of the observed patterns of population dynamics that you find documented in this literature.
3 Repeat the dodo exercise using a different model of population growth.
4 Consider which model best fits the growth of the wood bison population from the nucleus of 16 animals introduced into the Mackenzie Bison Sanctuary in northern Canada in 1963, based on the data presented by Larter et al. (2000, Fig. 2).
5 Introduce different patterns of environmental variability into any of the models given above or additional models that you have developed, and observe how the form of the output dynamics is affected.
6 Explore the effect of environmental variability further by making r rather than K vary randomly. This could be as a result of variability in the impact of predators on the survival rate. How does the outcome differ from the situation in which K is variable? How would you interpret your findings?

4

Structured population models

Age, size, or stage

Chapter features

Topics

Vital rates; Leslie matrix; Usher matrix; Lefkovitch matrix; life history stages; eigenvalue; stable age distribution; sensitivity analysis; projection versus prediction

4.1 Introduction

In Chapter 3, population growth was governed by a parameter r, representing the net difference between the population birth rate and death rate. But what is the meaning of a *population* birth rate? Only females reproduce, and only above a certain age. The death rate also differs between young, mature, and old animals, and perhaps between males and females. A population comprising entirely males, or immatures, would have a zero birth rate, whereas one consisting mostly of mature females would reproduce at a high rate. However, the population composition changes as a result of the births and deaths, thereby affecting the growth rate. In this chapter, you will learn how to derive the effective population growth rate inherent in the age- or stage-specific reproductive and survival rates, together with the resultant population structure. It provides the foundation for a demographic approach to population dynamics.

If the population that we are managing shows a disturbing trend, such as a persistent decline, what should be done to rectify this? Should we try to improve the survival of juveniles or the survival of adults? How much difference does an incremental (or proportional) change in each specific vital rate make to the overall population growth rate? Structured population models help decide where most effort should be directed. They can also be needed to evaluate the consequences of harvests selective for particular age or sex classes, such as trophy hunting largely for males, or fisheries where the mesh size of nets allows small fish to escape.

Representing the population size by a single number N is no longer adequate. We now need to know the numbers within each of the age or stage classes in the population. Similarly, instead of a single birth or death rate, we need to know the specific vital rates applicable to each class. Mathematically this leads to matrix algebra, which deals with sets of simultaneous equations governing the transitions within each of the classes constituting the population. A matrix is simply a list, array, or table of numbers arranged in a particular form. For example, for the population we need a list of numbers representing how many individuals there are in each class: for example, the number of newborns, yearlings, two year olds, etc. subdivided by sex. To project the transitions between these classes between one time step and the next, we need to arrange the reproductive and survival rates in a particular format. Note that matrix models are automatically discrete time models.

Models based on age (year) classes are appropriate mostly for mammals and birds, which have vital rates governed largely by age. Other organisms tend to show somewhat more flexible life histories than large vertebrates, dependent more on the stage or size reached than on age. For plants, variability in growth and hence size attained at a particular age can be huge. Trees can remain as seedlings on the forest floor for many years, awaiting a chance for a gap to open so that they can grow. Seeds may persist for even longer periods, germinating only when favorable conditions of soil moisture occur. For some invertebrates, fertile eggs can remain dormant for long periods until conditions are suitable for hatching. Fecundity can also vary greatly with size for reptiles and fishes as well as for plants.

You will start with the age-structured formulation applicable especially to mammals and birds. Thereafter you will be introduced to the modifications needed when reproductive outputs and survival chances depend more upon size or life history stage than on age. All of the basic models project exponential growth, once the age or stage structure has been stabilized, assuming that the set of vital rates remains constant. An important exercise will be to investigate the sensitivity of population growth rate to alterations in particular vital rates.

Matrix operations can be handled succinctly by many programming languages, such as True BASIC. The equivalent calculations can also be carried out very easily in a spreadsheet, the only complication being that you need separate columns of numbers to represent each class distinguished within the population. The appendix to this chapter, at the end of the book, provides guidance in formulating the spreadsheet models, whereas the accompanying CD contains the equivalent models written as computer programs.

4.2   Age-structured models

The information on vital rates needed for an age-structured population model comes from life history studies. Life history analysis was developed originally by actuaries to assess how the risk of death, as well as births of children, changed with age for human populations. This information forms the basis for life insurance premiums. The database was provided by birth and death registers. These document the birth of individuals as well as their parents, and later the year of their death, and hence how long they lived. From these records life tables were derived, estimating the probability of death (or its opposite, survival) and reproduction as a function of age.

A similar analysis based on "cohorts" born in the same year can be applied to animals. Birds can be fitted with leg rings (or bands) while in the nest, tags can be placed on fish, and ear notches or other marks on large mammals. The challenge is to find these animals when they die. Generally, a large sample of animals needs to be marked to get sufficient recoveries. Alternatively, composite life tables can be constructed from the age at death, identified from dentition or some other feature in a sample of found skulls, or for fish from growth rings in scales. However, for reliable estimates recoveries must be fairly complete and unbiased with respect to age, which is rarely true. Information is also required on the population growth rate over the period spanned, to eliminate a bias that would otherwise exist in the estimates. A third approach is to estimate annual survival rates from the age-class proportions in a population sample obtained at one specific point in time. This method is least reliable since the age-specific survival estimates will be distorted by annual variability in reproductive success as well as by a nonstationary population. Changes in the specific birth rate of females relative to age may be determined by monitoring cohorts. Alternatively, animals of known or identifiable age may be captured, or killed,

Table 4.1 A life history table constructed from skulls collected for a hypothetical ungulate producing twin offspring

Age class x	Number of skulls n_x	Survival frequency f_x	Survivorship l_x	Proportion of deaths d_x	Mortality rate q_x	Survival rate p_x	Fecundity rate m_x	Reproductive contribution $l_x m_x$
0	101	256	1.00	0.39	0.39	0.61	0	0
1	17	155	0.61	0.07	0.11	0.89	0	0
2	12	138	0.54	0.05	0.09	0.91	0.5	0.270
3	32	126	0.49	0.13	0.25	0.75	1.0	0.492
4	57	94	0.37	0.22	0.61	0.39	1.0	0.367
5	37	37	0.14	0.14	1.00	0.00	1.0	0.145
6	0	0	0	0				
							Sum	1.273

to determine their reproductive state, and perhaps their past reproductive history (e.g., from scars on the ovary). From the life history table (e.g., Table 4.1), the age-specific survival rates and reproductive outputs that go into the age-structured population model are obtained. Note that Table 4.1 represents the population increase per generation obtained by summing the values in the last ($l_x m_x$) column. The annual population increase is obtained by relating this rate to the generation time – see any population ecology text for details of the calculations.

An age-structured model calculates changes over time in the number of animals within each age class n_0, n_1, n_2, n_3, up to the maximum age reached. Time is typically incremented in steps of 1 year. Hence the first age class comprises animals aged between 0 and 1 year, the next class animals aged 1–2 years, and so on. It is important to recognize that animals automatically move on one class over each time step (Fig. 4.1). Hence individuals aged 0–1 years this year become 1–2 years the next year. The number of animals aged 1–2 years in this year is the number that were 0–1 years of age last year, multiplied by the survival rate between 0–1 and 1–2 years of age. These calculations are repeated for each age class, up to the oldest year class in the population. The survival rate of animals in this last class is zero, since they have reached the maximum longevity and drop out of the population. Thus, the number of age classes needed depends on the potential life span of the species.

Individuals enter the first age class via reproduction. Hence to obtain the numbers in the class n_0, the reproductive contributions of all of the age classes must be summed. This entails multiplying the number of females in each age class by the class-specific fecundity rate. By definition, fecundity is the number of *female* offspring produced by each female in a particular

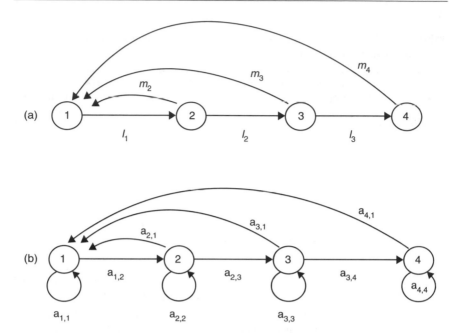

Figure 4.1   Life history loops comparing (a) an age-structured population with (b) a stage-structured population (from Silvertown and Doust 1993).

class. Specifically, the fecundity is the proportion of females conceiving (i.e., their fertility) multiplied by the litter or clutch size produced, and the female sex ratio of the progeny. In order to simplify calculations, the model is generally restricted to the female segment of the population. Males can be added later, assuming that they have no substantial influence on the reproductive output of the females. This assumption will fail to be true only when the adult sex ratio is so severely biased, for example, by hunting, that some females are not mated.

A small complication arises in the calculations from the stage in the annual cycle when the population is enumerated, relative to the time of birth or hatching. First, some of the adult females that were present the previous year may not survive to produce offspring in the current year. This means that the actual number of offspring contributed entails multiplying the age-specific fecundity additionally by the survival rate of females in this class. Furthermore, the fecundity assessment is based on age in the previous year, for example, females in the age class 0–1 years have a fecundity determined by the offspring they produce in the following year when aged 1–2 years. Second, unless the population census is done immediately after the birth pulse, before any offspring mortality has occurred, the survival rate of the offspring between birth and the time of the census must also be taken into account. Hence the fecundity values used to calculate reproductive contributions to the first age class 0–1 years are actually a composite product of the number of female offspring produced, the age-specific survival of mothers, and the survival rate of the offspring between birth (or hatching) and when the census is conducted.

Having assigned the age-specific vital rates, the model establishes the population growth rate that is generated. The growth rate may fluctuate initially as a consequence of the starting age structure that was assigned. However, eventually the age structure stabilizes and the growth rate becomes constant (Fig. 4.2). Set up a spreadsheet model to explore this pattern following the guidance given in Appendix 4.1, or run the True BASIC program called AGEINTR. Note how the eventual population growth rate is the same, whatever the initial population structure.

4.2.1 Sensitivity or elasticity analysis

The next task is to establish the relative contribution of each vital rate toward the overall population growth rate. This entails conducting either a *sensitivity* analysis or an *elasticity* analysis. The difference between them arises from how the vital rates are adjusted. For a sensitivity assessment, each rate is altered by the same *amount*, for example, a reduction in juvenile survival by 0.1 from 0.5 to 0.4 per year is compared with the effect of lowering adult survival similarly from 0.9 to 0.8 per year, or fecundity from 1.0 to 0.9 per year. For an elasticity assessment, the rates being compared are adjusted by the same *proportion*, for example, reducing juvenile survival by 10% from 0.5 to 0.45 versus lowering adult survival from 0.9 to 0.81. Which comparison is more appropriate depends on the magnitudes of the respective rates. For example, for a bird laying up to 20 eggs in a clutch, it would be trivial to compare reducing fecundity by an amount of 0.1, from 10 to 9.9, with the consequences of a reduction in adult survival from 0.9 to 0.8. Hence an elasticity analysis would be most appropriate. On the other hand, when comparing the effects of changes in survival, adjustments by the same amount might be more meaningful in order to avoid the dependence of elasticity on the effective mean value. A further consideration is whether it is appropriate to compare the effect of changes in survival of just a single age class, for example, juveniles, with the consequences of similar adjustments across a wider range of adult classes.

In general, for large, long-lived mammals, the population growth rate is most sensitive to changes in adult survival, because adults constitute half or more of the population. The situation is rather different for short-lived animals with high fecundity, because the young animals then form the

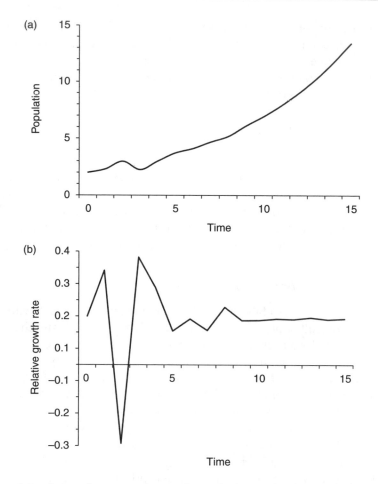

Figure 4.2   Output from age-structured population model showing (a) change in population size over time and (b) change in relative growth rate over time, from starting population of one adult plus one juvenile.

majority of the population. The findings from an elasticity perspective could differ if the survival rate of these young is basically quite low. Undertake a preliminary exploration of these patterns by adjusting the age-specific birth and death rates entered into your elementary age-structured model.

Changing the adult survival by the same amount or proportion as juvenile survival or fecundity tells you how much difference such a change *could* make, if it occurred, and so is called a *prospective* analysis. Making changes based on the actual range of variability that has occurred in the past tells you to what extent fluctuations in population growth rate have been affected by variability in different vital rates, and so is termed a *retrospective* analysis. See Gaillard, Festa-Bianchet, and Yoccoz (1998) and Caswell (2000) for deeper discussion of these perspectives. Heppell, Caswell, and Crowder (2000) describe how elasticity analysis can be applied to simplified models for a number of mammal populations where only minimal demographic data are available.

4.2.2 Leslie matrix

The set of simultaneous calculations involved in projecting the future population growth from a combination of vital rates together with the starting age structure is readily handled by matrix algebra. This requires arranging the vital rates in the form of a two-dimensional table or array, with the set of fecundity rates f_j in the top row and the class-specific survival rates s_i along the off-diagonal. The *is* label specific age classes with x representing the oldest age class:

$$\begin{pmatrix} f_0 & f_1 & f_2 \ldots f_x \\ s_0 & 0 & 0 \ldots 0 \\ 0 & s_1 & 0 \ldots 0 \\ & \cdots & \\ 0 & 0 & 0 \ldots s_{x-1} 0 \end{pmatrix}$$

Correspondingly, the numbers of animals n_i in each age class are arranged as a column or a one-dimensional array (also called a column vector):

$$\begin{pmatrix} n_0 \\ n_1 \\ n_2 \\ \vdots \\ n_x \end{pmatrix}$$

This follows from the rules for matrix multiplication. To get the numbers in a particular age class in the next time step, the starting numbers in each cell in the column are multiplied by the set of numbers in the corresponding row of the square array. For instance, to get the number in age class n_0 next year, we calculate $n_0 \times f_0 + n_1 \times f_1 + \cdots + n_x \times f_x$. Note that this is exactly equivalent to what you did in your elementary spreadsheet model. For remaining age classes, the calculation is simpler because there is only one transition rate in each row representing the survival rate from each age class into the next. If these survival rates had been placed along the *diagonal* of the array, surviving animals would remain in the same age class, which cannot happen since age must increase over each time step.

The particular form of matrix depicted above is called the *Leslie* matrix, after the ecologist who first used it to project population growth. In the formalism of matrix algebra, we can write $N_{t+1} = L * N_t$, where N represents the population vector and L represents the Leslie matrix. Correspondingly, in True BASIC, the matrix multiplication can be performed using a one-line statement. However, it is important to recognize that all of the operations are performed on the numbers in each age class in the preceding year. Hence the values for "fecundity" in the top row must incorporate the necessary corrections for the survival of mothers and the offspring they produce between successive counts. They are better interpreted as the *recruitment* into the first class emanating from the animals in each age class within the population in the previous year. The program SMATRIX on the accompanying CD illustrates how simple matrix operations can be conducted in True BASIC. The program LESLIE uses a more elaborate Leslie matrix to generate and graph the dynamics of an age-structured population. It also allows any of the vital rates to be changed proportionately for an elasticity analysis.

In matrix algebra, the population growth rate inherent in the set of vital rates as a multiplication factor is formally termed the dominant *eigenvalue* of the matrix, symbolized by λ. After the age structure has stabilized, the population size the next year can be obtained simply by multiplying each age class by this single number because at this stage all age classes are increasing at the same rate. The corresponding stable age distribution is termed the *eigenvector*. Mathematicians have analytic techniques for extracting the values of λ and its associated eigenvector. With the aid of a computer, you can obtain the same answer numerically. Note that the Leslie matrix generates geometric population growth because the fecundity and survival rates remain constant.

Instead of the matrix multiplication, a FOR ... NEXT loop could be used to perform the relevant calculations. This allows somewhat greater flexibility. For example, instead of incorporating the relevant survival rates into the effective values for fecundity, you could calculate the reproductive contribution after deriving the number of surviving females in each age class. The program AGEINTR uses this approach.

4.2.3   Usher matrix

For many organisms, survival rates remains fairly constant once maturity has been reached, at least in relation to the inevitable errors in their estimation from small samples. Thus, it seems hardly necessary to type in the same survival rate over and over again, across all adult classes. Why not just represent the adult survival rate in one final age class? This is what is done in an Usher matrix, named after another ecologist. In matrix notation, it takes this form for four age classes:

$$
\begin{pmatrix}
f_0 & f_1 & f_2 & f_3 \\
s_0 & 0 & 0 & 0 \\
0 & s_1 & 0 & 0 \\
0 & 0 & s_2 & s_3
\end{pmatrix}
$$

There is now a nonzero value for survival in the final diagonal slot (s_3 in the above example). This means that individuals can survive in this last age class indefinitely. Although unrealistic, this simplification makes little difference to the projected population growth rate because very few individuals live long enough to die of old age for most organisms. Moreover, the potential longevity of a species in the wild is usually unknown anyway.

Explore the effect of representing adult survival in this way by comparing the population growth rate generated by the program LESLIE with that of the program USHER (both on your CD), for the same vital rates (or formulate spreadsheet models to make this comparison, following the guidelines given in Appendix 4.1).

Structured population models could be used to establish the maximum population growth rate that a particular species could exhibit under ideal conditions, from its life history features: that is, the age at first reproduction, litter or clutch size, and the interval between births and potential life span. Mortality can be set at or close to zero, except possibly for the first age class. The reproductive potential largely determines the potential population growth rate, that is, the value of r_0 for the logistic or equivalent descriptive equation.

4.3 Stage-structured models

A shortcoming of the Leslie matrix is that it represents a set of almost identical survival rates across the adult classes, while compressing various processes contributing to recruitment into one outcome for fecundity. Furthermore, for many organisms, the effective vital rates depend more on the life history stage or size reached rather than age. The period required to pass through these stages can be variable, depending on conditions. Even for mammals, early growth affects the time taken to reach adult size and hence the age at which reproduction commences. Last, why count new individuals only after birth or hatching when life begins at conception?

Embryos within the uterus, egg, or seed are strictly part of the population despite being invisible.

This leads to a different form of matrix, structured into stage classes rather than age classes. Its original formulation was by an ecologist named Lefkovitch, for insect populations, and therefore it goes by his name. For an insect the stages could represent the sequence of larval instars or nymphs through which individuals pass before they reach the adult or reproductive stage. The appropriate time steps would then be days or weeks, rather than years. The rate of progression through these stages may be influenced by both temperature and food supply. Hence while some individuals may remain in the same stage others may have grown toward the next stage over the same period. Eggs may also remain in a state of dormancy for a variable period. To accommodate this, entries are now needed in the diagonal cells as well as the off-diagonal ones of the transition matrix:

$$\begin{pmatrix} a_{0,0} & 0 & c_2 & c_3 \\ a_{0,1} & a_{1,1} & 0 & 0 \\ 0 & a_{1,2} & a_{2,2} & 0 \\ 0 & 0 & a_{2,3} & a_{3,3} \end{pmatrix}$$

Thought needs to be given to what these parameter values represent, aided by the life history loops represented in Fig. 4.1b. At the end of each time step, individuals can either (i) be still in the same stage, (ii) have progressed to the next stage, or (iii) have died. Each of these fates has a certain probability, assessed by the proportion of individuals that incur it. The probability of remaining in the same stage is given by the entries $a_{i,i}$ along the diagonal. Hence the entry $a_{0,0}$ is not the fecundity of the first stage, which represents fertile eggs, but rather the proportion of eggs that remain unhatched but still viable at the end of the time step. The numbers $a_{i,j}$ along the off-diagonal represent the probability of growing into the next stage, for example, $a_{0,1}$ represents the proportion of eggs that have hatched into first-stage larvae. Hence the survival rate is the sum of $a_{i,i}$ and $a_{i,j}$ for each stage class, that is, of the two values paired vertically in the table. The crucial condition is that these two proportional rates cannot add up to more than 1.0. The difference between their sum and 1.0 is the mortality rate for the stage.

The growth toward maturity of reptiles and fishes also depends on environmental conditions, with reproductive output and survival chances governed more by size than by age. Moreover, reptiles like tortoises live to enormous ages, so that an age-structured matrix of vital rates would be rather cumbersome. In a model developed for sea turtles, seven life history stages were distinguished: (i) first-year recruits (eggs or hatchlings),

(ii) small juveniles, (iii) large juveniles, (iv) subadults, (v) novice breeders, (vi) first-year remigrants, and (vii) mature breeders (Crouse, Crowder, and Caswell 1987). A stage-classified model might be appropriate even when growth is determinate, especially when age estimation is unreliable, for example, such models have been applied to small mammals (Sauer and Slade 1987). The Usher matrix is in effect a stage model for a large vertebrate, separating immature stages from a single adult stage. The key difference is that mammals and birds cannot persist in a juvenile stage. For very long-lived mammals like elephants, a time step of more than a year might be appropriate.

Refer to Appendix 4.2 for guidance in formulating a stage model for a sea turtle population and then using this model to address a specific conservation problem. For plant populations, a simple stage matrix could represent just three life history stages: seeds, vegetative plants, and reproductive plants. However, in contrast to animals, plants can also undergo stage reversals. A tree felled by wind-throw (or by an elephant) could revert to a nonreproductive stage through coppice regrowth from the stump. New plants may also be produced through tillering or other forms of vegetative replication, as well as through seed germination. Hence the transition matrix for a plant population requires entries in additional cells:

$$\begin{pmatrix} a_{s,s} & 0 & a_{r,s} \\ a_{s,v} & a_{v,v} & a_{r,v} \\ 0 & a_{v,r} & a_{r,r} \end{pmatrix}$$

where $a_{s,s}$ is the probability that seeds remain as seeds; $a_{v,v}$, the probability that vegetative plants remain vegetative; $a_{r,r}$, the probability that reproductively mature plants remain alive as mature plants; $a_{s,v}$, the proportion of seeds germinating to become vegetative plants; $a_{v,r}$, the proportion of vegetative plants growing into reproductively mature plants; $a_{r,s}$, the number of seeds produced by each reproductively mature plant; $a_{r,v}$, the number of new vegetative plants produced by each reproductively mature plant, including mature plants reverting to the vegetative state.

By definition of this stage, vegetative individuals cannot produce seeds, so the cell representing their fecundity must remain zero. The cell in the lower left corner is zero because it would represent a transition from seeds across to mature plants in one time step. The iteration period should be set to preclude this.

Explore how the spreadsheet model for a turtle population outlined in Appendix 4.2 might be modified to represent a tree population by adding these additional stage transitions. Annual seed production from a mature tree could be enormous, and coupled with a long life span might seem to predispose tree populations to a huge potential rate of increase. However, in most tree populations, only a small fraction of seeds survive the attack of parasites and granivores to enter the soil in a viable state, and very few of the seedlings that germinate survive to become mature trees. Explore the combinations of vital rates that produce a reasonable rate of growth for a tree population. An iteration period of several years would be appropriate. Compare the output of your spreadsheet model with that of the programs LEVKOVIT and STAGESTR on the CD.

4.4   Projection versus prediction

The tables or matrices containing sets of vital rates for age or stage classes are commonly referred to as *projection* matrices. They merely project the population growth rate that would be generated if these vital rates remained constant for a sufficient period. In the real world, conditions vary and survival and reproductive outputs are affected by fluctuations in weather as well as density feedbacks. The consequent variation in vital rates means that a stable age distribution is never attained, except perhaps during periods when a growing population is so far below the food ceiling such that resource limitations are ineffective. To *predict* what the actual population size will be at some future time, you would need to take into account the effects of these perturbations by making the survival rates functions of resource availability and population density. Ways in which this could be done will be covered in Chapter 6.

Other factors besides age, size, and life history stage can also influence survival and reproduction, for example, social status, early experience, or even the surrounding neighborhood. This can lead into finer partitioning of the population structure, ending ultimately in *individual-based* models (DeAngelis and Gross 1992). However, the computational demands of such models can be huge.

4.5   Overview

What you should have learned from this chapter is the following:

1 Populations may be structured into classes of individuals differing in their reproductive and survival rates.
2 Projection matrices form a convenient tool for calculating the population growth rate inherent in a set of specific vital rates and are easily represented in a spreadsheet.

3 The Leslie matrix emphasizes age-related changes in survival and fecundity among mature classes.
4 The Usher matrix is advantageous when survival and fecundity change little with age within the adult class.
5 The Lefkovitch matrix accommodates indeterminate growth between stage classes.
6 A sensitivity or elasticity analysis establishes the relative effect on population growth of changes in specific vital rates.

Recommended supporting reading

The definitive book on matrix population models is Caswell (2001). Simpler accounts are presented by Gotelli (1995), Vandermeer and Goldberg (2003), and Manly (1990). The article by Groenendael, de Kroon, and Caswell (1988) is a useful overview.

Programs on the accompanying CD

AGEINTR; SMATRIX; LESLIE; USHER; LEFKOVIT; STAGEST

Exercises

1 Compare the maximum population growth rates that could be manifested by (i) a typical African antelope, producing a single young annually and first conceiving at 2 years of age, (ii) an American moose or white-tailed deer, producing twins or even triplets annually, (iii) a small antelope producing a single young annually but first conceiving at 1 year of age, and (iv) a warthog, first conceiving at 2 years of age and producing 3–4 piglets annually thereafter.
2 Incorporate appropriate vital rates into a structured population model to represent an animal or plant species of your choosing. Undertake a sensitivity analysis of the effects of changes in population parameters on the inherent rate of population growth. Rank population parameters in the order of their importance. Compare your findings with those of a classmate considering a different organism. If there are differences in your conclusions, try to work out why.
3 Compare the output from your spreadsheet model with that from one of the True BASIC programs supplied on disk, for the same vital rates. If you get a discrepancy, try to establish why.

4 Find a paper in the literature presenting age- or stage-specific survival
and fecundity rates for any organism, and enter these rates either into a
spreadsheet or into one of the True BASIC programs supplied on the CD.
5 Consider how to incorporate the male segment into a structured population model, for example, to estimate the number of trophy-age males
that can be produced annually to support a hunting quota.

5

Consumer–resource models

Population interactions

Chapter features

Topics

Functional or intake response; numerical response; biomass gain function; delayed density dependence; herbivore–vegetation interactions; predator–prey oscillations; paradox of enrichment; resource dependence; interference competition

5.1 Introduction

In Chapter 3, the population growth rate was expressed as a function of the population density relative to some arbitrary "carrying capacity." What determines this maximum abundance level was left mysterious. The kinds of models that will be developed in the current chapter relate population growth directly to food resources and also represent the impact of predators on the growth of their prey. Hence we now represent the interactive dynamics of coupled populations, either a herbivore dependent on plants for food or a carnivore predating some herbivore species. This enables us to see directly how the abundance of the consumer population emerges directly as a function of food abundance, quality, and accessibility. At the same time, through capturing and consuming plant parts or prey animals, consumers affect the abundance of the resources upon which they depend. The lags inherent in this interaction tend to generate oscillatory dynamics rather than a smooth approach toward some carrying capacity level.

The foundations for interactive population dynamics were laid by Lotka (1925) and Volterra (1926) in the form of the Lotka–Volterra equations. They were trying to explain why some species show regular oscillations in abundance, notably lemming, voles, snowshoe hare, and associated species in the far north of Europe and Canada. Some ungulate populations also show "irruptive" dynamics, increasing to high abundance levels, then crashing as a result of lack of food. It has been suggested that elephants and trees might oscillate indefinitely in abundance because of the severe damage that elephants can impose on woody plants in the course of their feeding (Caughley 1976b). This has led to controversial culling programs to restrict the abundance of elephant within protected areas. However, most populations do not display persistent oscillations, and the challenge is to explain why these different patterns exist.

Basic considerations relate to whether the consumer population merely responds to changes in its food resources, or interacts with these resources to change their supply rate. Also, to what extent do the consumers interfere with each others' food gains? What are the consequences of seasonal and annual variability in food production and availability for the consumer population dynamics? How is overgrazing, or overexploitation of resources in general, manifested?

5.2 Coupling population equations

To represent the joint dynamics of both consumer and resource populations, we need two equations linked through the trophic interaction. In schematic outline,

$$dX/XdT = R(X) - I(X, Y)$$

and

$$dY/Ydt = G(X, Y) - M(Y, Z),$$

where X represents the resource, Y the consumer, and R the production function for the resource, which depends only on the abundance X of the resource. I is the intake function, representing the amount of the resource consumed, which depends both on the abundance of the resource and on the abundance of the consumer. G represents the function governing the gain in abundance of the consumer as a result of the resources consumed, which is likewise potentially dependent on the abundance of both populations. M represents metabolic and mortality losses incurred by the consumer population, through physiological attrition, predation, senescence, and other extraneous causes. The mortality loss is affected also by the abundance Z of other populations, such as predators and parasites. Note that the dynamical equations have been expressed *relative* to the respective abundances of the two populations, as shown by the appearance of X and Y in the denominators on the left-hand side.

In theoretical ecology, the intake function I is usually called the *functional response* of the consumer to food abundance. This name is a little misleading because there are four functional responses to be considered in the above pair of equations. The equation expressing the consumer population dynamics as the difference between food gains G and losses M is commonly termed the *numerical response*. This is also somewhat misleading because in coupling the population dynamics, we need to work in biomass units. The increase in the abundance of consumers depends not only on the number of food items eaten, but also on their size. Note that a birth rate is no longer explicitly represented. The increase in the consumer population is an outcome of the amount of food gained, which presumably affects how many offspring are produced and how successfully they survive.

The dynamics manifested by the interactive populations depend on the functional form of the relationships and on the parameter values in these functions. While textbooks generally emphasize predator–prey interactions, the herbivore–plant interaction is perhaps more fundamental. Hence the focus in this chapter will be on herbivore–vegetation relationships. We can assume that plant populations merely respond to the supply of the resources they need – sunlight, rainfall, atmospheric carbon dioxide, and mineral nutrients obtained from the soil – without affecting these supply rates. However, herbivores consuming plants and predators consuming herbivores affect the abundance of their resources. Tri-trophic level models coupling plants, herbivores, and carnivores suggest that predation can even affect the amount of vegetation persisting.

5.3 Simple interactive model

5.3.1 Linear intake function

The simplest intake function to consider is a linear dependency of the herbivore population gain on the amount of vegetation consumed. Each time a herbivore contacts a plant, some amount of the plant is consumed and converted into herbivore biomass. The food intake rate will thus be a function of the area searched per unit time per herbivore s, the amount of vegetation within this area V, and the fraction of the vegetation eaten and digested. Assume further that metabolic and mortality losses occur at a constant proportional rate m. We then have

$$\Delta H/H\Delta t = csV - m, \tag{5.1}$$

where H represent the herbivore biomass density.

With this formulation, the herbivore population grows exponentially without limit if $csV > m$, and declines exponentially if $csV < m$. However, the value of V depends on how much food is consumed, and hence is changed as the herbivore population grows. We need to turn our attention next to the production function for the vegetation.

It seems reasonable to assume that, in the absence of herbivores, the growth of a plant population will be affected by competition for resources within the space that it occupies. Under these conditions, the plant biomass should increase at a diminishing rate until all available space has been taken up, and no further growth can take place until some plants die. In these circumstances, the logistic equation should be an adequate description of the intrinsic plant population dynamics; that is,

$$\Delta V/V\Delta t = r_V(1 - V/K_V), \tag{5.2}$$

where V again represents vegetation biomass density, r_V is the maximum growth rate of the vegetation, and K_V the maximum biomass level attained by the vegetation. To avoid seasonal complications, we will adopt an annual time step, so that r_V represents the proportional increase in plant biomass from one year to the next. In the absence of herbivores (or any other disturbance), plants should remain at their zero growth level K_V.

However, by eating some of the plant material, the herbivores reduce the plant biomass below this level. Hence an additional term needs to be added to account for the amount consumed, that is, eqn (5.2) becomes

$$\Delta V/V\Delta t = r_V(1 - V/K_V) - sH. \tag{5.3}$$

Hence the larger the herbivore biomass, the greater the reduction in the amount of vegetation remaining after consumption. However, having been reduced below the carrying capacity sustained by its resources, the plant population can recover at a rate governed by its logistic growth.

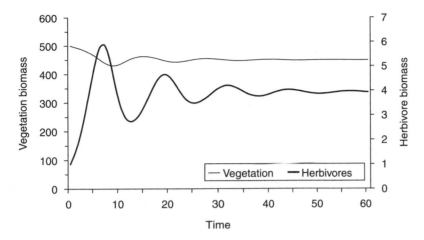

Figure 5.1   Output of the interactive herbivore–vegetation model with a linear intake response for these parameter values: $r_V = 0.5$, $K_V = 500\,\mathrm{g\,m^{-2}}$, $s = 0.02\,\mathrm{year^{-1}}$, $c = 0.5$, $m = 4.5\,\mathrm{year^{-1}}$.

Set up this model in a spreadsheet following the guidelines given in Appendix 5.1, including the parameter values suggested. You should find that both the herbivore and plant populations show initial oscillations in abundance if you allow the herbivore population to increase from some low level (Fig. 5.1). Oscillations result from delays in the response of the herbivore population to the changing food availability, and in the subsequent recovery of the vegetation following a decline in the herbivores. The plant population eventually stabilizes at some level slightly below its maximum potential. At the joint equilibrium, the amount of vegetation is just sufficient for the herbivore to gain enough food to balance its metabolic and mortality losses.

However, the output dynamics depend critically on the parameter values chosen. An increase in the productivity of the vegetation, through raising either r_V or K_V, amplifies the oscillations, largely through raising the herbivore biomass that can be supported and thereby the vegetation impact. This finding is known as "the paradox of enrichment" (Rosenzweig 1971). Increasing slightly the value of s, governing the searching efficiency of the herbivores, or c, raising the nutritional value of the food, increases the potential growth rate of the herbivore population, and hence the amplitude of the oscillations. Too great an increase in these parameters leads to a crash by the herbivore population into negative numbers. Lowering these parameters more than slightly stops the herbivore population from increasing. Elevating the value of m, that is, the metabolic and mortality costs incurred by the herbivores, slows the growth rate of the herbivore population and results in a lower herbivore density at equilibrium. The herbivore population declines exponentially if these biomass losses outweigh the value of the food gained.

5.3.2 Saturating intake function

It is somewhat unrealistic to assume that the intake rate I of the herbivores increases without limit as the amount of vegetation becomes greater. Eventually the herbivores will be consuming food as fast as they can chew, swallow, and digest it. This constraint imposed by handling time sets an upper limit to the food intake rate, which is likely to be reached asymptotically. Following a bout of foraging, herbivores cannot resume searching for more food until the food already obtained has been digested. This leads to a "Type II" functional response, following the terminology of Holling (1965):

$$I = sV/(1 + sV/h), \tag{5.4}$$

where $h =$ digestion time required per unit mass of food. When food is very abundant, this equation reduces to $I = 1/h$, that is, the maximum intake rate is inversely related to the time required to digest the food.

You should recognize eqn (5.4) as taking the basic form of a rectangular hyperbola, that is, the food intake rate increases asymptotically toward its maximum value as the amount of vegetation available as food increases (the Beverton–Holt equation for population dynamics had this form). Since all we need to do is describe the form of the intake function, without being concerned about the specific mechanisms, we can express eqn (5.4) more simply as

$$I = aF/(b + F), \tag{5.5}$$

where F represents the amount of vegetation that is effectively available as food (which is likely to be somewhat less than the total amount of vegetation V), a governs the maximum intake rate when food is very abundant, and b determines how steeply the food intake rate declines toward low food abundance levels. Specifically, b is the amount of food at which the intake rate reaches half of its maximum. Equation (5.5) is known as the Michaelis–Menten function.

Accordingly, the vegetation dynamics equation becomes

$$\Delta V/\Delta t = r_V V(1 - V/K_V) - aHF/(b + F) \qquad (5.6)$$

while the herbivore dynamics is now

$$\Delta H/H\Delta t = caF/(b + F) - m. \qquad (5.7)$$

Caughley (1976a) adopted this model, due originally to Rosenzweig and MacArthur (1963), to represent the dynamics of herbivore–vegetation systems, with one small difference; he chose an exponentially saturating intake response. However, the hyperbolic formulation has since become most generally used because it is more easily interpreted.

Incorporate the changed intake response into a new spreadsheet (see Appendix 5.2). With a suitable choice of parameter values, you should observe oscillations decaying toward equilibrium populations of both herbivores and vegetation (Fig. 5.2). Raising the value of the half-saturation level b slows the initial population growth rate at low density, which could allow to an asymptotic approach toward an equilibrium level as produced by logistic growth. Lowering b, or raising the conversion coefficient c, amplifies the oscillations, which may become sustained. Lowering m, controlling the metabolic and mortality costs incurred by the herbivores, has a similar effect. However, if you adjust these parameters by more than a small amount, the output heads into negative numbers. In other words, the dynamics are highly sensitive to how efficiently the herbivores feed on sparse vegetation and on the nutritional value of the vegetation consumed. If the herbivores still gain sufficient nutrition to maintain their population even when little vegetation remains, oscillations are promoted because the herbivore population responds little to the diminishing vegetation until not much remains. Following the crash in the herbivore population due to inadequate food, the vegetation recovers. Small increases in the mortality loss, for example, through elevated predation, can dampen or even suppress the oscillatory tendency.

When the relative growth rate of the herbivore population is plotted against the herbivore biomass density for a situation showing dampened oscillations, a convex trend spiraling in toward the zero growth level is shown (Fig. 5.3). This should remind you of the pattern observed when an explicit time delay for the density feedback was introduced into the logistic equation in Chapter 3. The lag inherent in the herbivore–vegetation interaction is revealed by plotting the herbivore growth rate against the

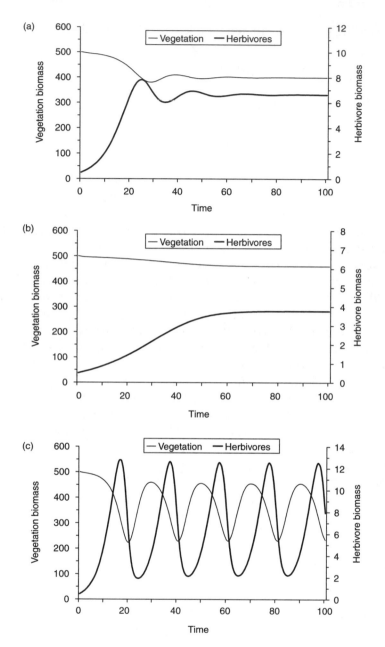

Figure 5.2 Output of the interactive herbivore–vegetation model with a saturating intake response: (a) $a = 10\,\text{year}^{-1}$, $b = 100\,\text{g}\,\text{m}^{-2}$, $c = 0.5$, $m = 4.0\,\text{year}^{-1}$; (b) $a = 10\,\text{year}^{-1}$, $b = 115\,\text{g}\,\text{m}^{-2}$, $c = 0.5$, $m = 4.0\,\text{year}^{-1}$; (c) $a = 10\,\text{year}^{-1}$, $b = 85\,\text{g}\,\text{m}^{-2}$, $c = 0.5$, $m = 4.0\,\text{year}^{-1}$.

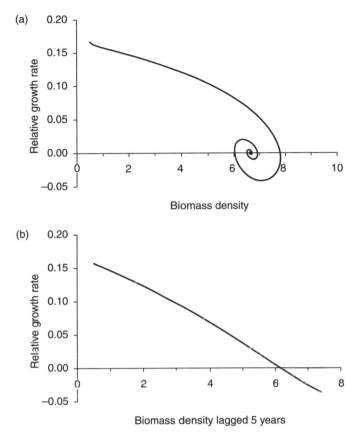

Figure 5.3   Density dependence in herbivore population growth rate generated by the model with saturating intake response and parameter values as in Fig. 5.2a, plotted against (a) density in current year and (b) density 5 years earlier.

herbivore biomass density several years previously. With an appropriate lag time, you should find that the delayed density dependence becomes more or less linear. The delay arises because the density feedback occurs indirectly via the impact of the growing herbivore population on the food resources that later remain available.

The dependence of the growth dynamics of the herbivore population on food availability can be revealed by plotting the relative growth rate as a function of the amount of vegetation, using eqn (5.5). This reveals the minimum amount of vegetation needed to sustain the population (Fig. 5.4). If the vegetation biomass exceeds this level, the herbivore population grows, otherwise it declines. The location of this breakeven point is determined by the values chosen for b, c, and m.

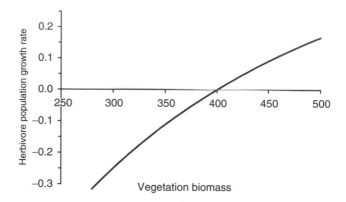

Figure 5.4   Herbivore population growth rate plotted against the vegetation biomass, from eqn (5.5), with parameter values set as in Fig. 5.2a.

5.4   Incorporating competitive interference

Equation (5.7) describing the growth of the herbivore population contains no density-dependent term. Nevertheless, the growth rate of the herbivore population does slow toward some zero growth level, with an appropriate choice of parameter values. The density feedback arises indirectly through the interaction between the herbivore population and its food resources, that is, more herbivores means less vegetation remains to counterbalance the mortality losses. For large herbivores it is reasonable to assume that there is little direct interference between animals feeding close together, because vegetation resources are so widely distributed. If prevented from feeding in a patch because it is occupied by another animal, a herbivore need move only a few steps to find another patch offering food. Hence the food intake rate is affected very little by the size of the population.

This assumption may not hold for other kinds of animals feeding on food types that are more locally concentrated. Pecking orders are well known for many birds. The bird doing the pecking loses some feeding time, as also does the bird that moves away. The greater the density of birds in the vicinity, the more time diverted from feeding to competitive interactions.

Equation (5.7) can be modified to incorporate a factor dependent on the consumer population density affecting the rate of food gain, following DeAngelis, Goldstein, and O'Neil (1975):

$$\Delta H / H \Delta t = caF/(b + F + dH) - m. \tag{5.8}$$

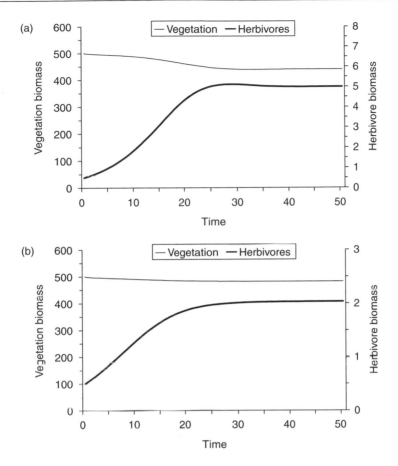

Figure 5.5 Output of interactive herbivore–vegetation model including an interference term with basic parameter values set as in Fig. 5.2a: (a) $d = 2$ and (b) $d = 10$.

The coefficient d sets the strength of the interference effect. Explore the consequences of allowing for competitive interference in your spreadsheet (see Appendix 5.3), noting in particular how a sufficiently large value of d dampens the oscillatory tendency (Fig. 5.5). The interference term introduces *direct* density dependence, in addition to the delayed density dependence generated by the consumer–resource interaction.

With sufficiently strong interference, the output dynamics closely replicate those generated by the logistic or other descriptive equations. The important difference is that the zero growth level of the herbivore population emerges from the settings of the parameters governing foraging efficiency, rather than being assigned some arbitrary level. Note further that if only a small fraction of the vegetation is available for consumption, that is, F is much less than V, the impact of consumption on the vegetation dynamics becomes greatly reduced, and the herbivore population becomes food-limited even though much vegetation still remains. These circumstances also act to stabilize the dynamics of the two populations.

5.5   Ratio-dependent intake response and time frames

It is reasonable to assume that the dynamics of the consumer population respond directly to the prevailing food availability in the short term. However, when representing dynamics over discrete time steps of a year, the amount of food that is effectively available to support these animals over the annual cycle will depend on the abundance of the consumer population. This leads to a "ratio-dependent" formulation of the intake response, that is, the food share available to each consumer is the amount of food divided by the number or biomass of consumers (see Berryman 1992 for a discussion). Hence, assuming the consumers to be herbivores, eqn (5.5) becomes

$$I = a(F/H)/\{b + (F/H)\}. \tag{5.9}$$

The ratio-dependent formulation of the intake (or functional) response acts to stabilize the consumer–resource interaction because the food gain becomes directly dependent on the biomass density of the consumer population. However, whether this representation is appropriate has remained contentious (see Abrams and Ginzburg 2000). It depends basically on the time perspective adopted in the model.

5.6   Accommodating environmental variability

The amount of edible vegetation produced is likely to vary from year to year depending on the weather conditions, in particular the precipitation received. Hence the potential vegetation biomass K_V supporting the herbivore population is a variable influenced by the state of the environment at the time. If you incorporate annual variability in food availability into your spreadsheet, following the guidelines in Appendix 5.4, you should find that this acts to destabilize the interactive dynamics of the system, especially if the vegetation is reduced to a level where the herbivore population growth becomes negative, even when few herbivores remain.

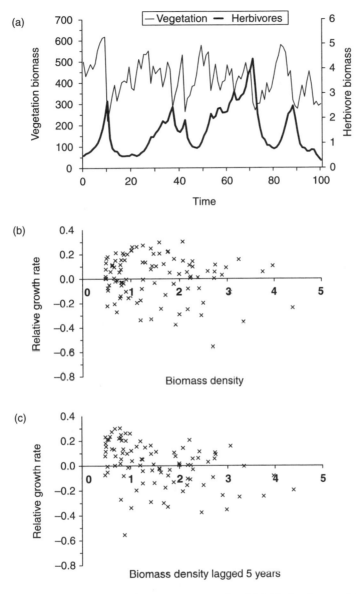

Figure 5.6 Dynamical output for circumstances in which K_V for the vegetation fluctuates randomly from year to year. (a) Changes in biomass over time, (b) direct density dependence in the current year, and (c) lagged dependence on density 5 years earlier.

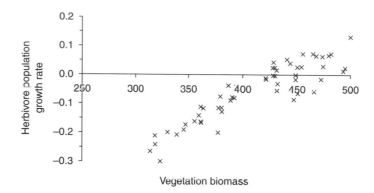

Figure 5.7 Herbivore biomass growth rate plotted against the vegetation biomass, for circumstances with fluctuating K_V and strong interference competition ($d = 10$).

Due to the nonlinear relationship between herbivore population growth and vegetation resources (Fig. 5.4), the herbivore population loses more biomass during the bad years than it gains from the good years. Moreover, the basic reproductive potential of the species may cap the recovery rate of the herbivore population when food is plentiful.

With wide environmental fluctuations, plotting the relationship between herbivore population growth and herbivore density will reveal little indication of direct density dependence, especially if circumstances when the herbivore population was very low are not represented (Fig. 5.6). However, allowing for a delay in the density feedback reveals its negative influence on population growth. What does this mean for attempts to detect density dependence under real-world conditions?

When the herbivore growth rate is plotted against the prevailing vegetation biomass, the curvilinear relationship remains clearly evident. This shows that the dynamics of the consumer population depend directly on resource availability. The herbivore biomass merely modifies the amount of vegetation remaining available. However, the resource relationship becomes blurred if there is strong interference competition as well as the resource interaction (Fig. 5.7).

The consequences of environmental variability can be explored through a model developed to establish the optimal stocking density for a livestock ranch. The aim is to establish the stocking level that gives the maximum sustainable production of meat from these animals, taking into account the dynamics of the vegetation resources. Consult Appendix 5.5 for guidance in setting up such a model in a spreadsheet. Experiment with various stocking levels, and observe what happens to the summed gain in animal production over some reasonably long time horizon, say 30 years, and also to the vegetation biomass. Identify the state of the vegetation giving the maximum return in terms of meat or animals produced. Having run the model initially in a constant environment, investigate how the optimal stocking level would need to be adjusted when the potential amount of vegetation varies between years as a function of rainfall. A more elaborate herbivore model distinguishing the growth and population dynamics of the grass resource will be developed in Chapter 11.

5.7   Overview

Interactive consumer–resource models reveal how the dynamics and abundance level of a herbivore or predator population depend on its interaction with its food resources. Instead of assigning an arbitrary population growth rate and carrying capacity, these emerge from factors governing food consumption and the response of the food resource to exploitation. The appropriate population currency shifts from numbers to biomass density, especially for the herbivore–plant interaction. From these models you should have learned the following:

1 The range of parameter values enabling persistence of the consumer–resource interaction is very narrow.
2 The saturation biomass level for a herbivore population depends on how efficiently the herbivores consume and digest the vegetation components providing their food resource.
3 The tendency toward instability is highly sensitive to the nutritional value of the food consumed and how efficiently the consumers gain food when resources are depressed.
4 Density dependence can arise indirectly through the consumer–resource interaction.
5 Lags in the interaction between the two populations can lead to coupled oscillations rather than a smooth approach toward equilibrium densities.
6 Both interference competition and additional mortality through predation dampen the oscillatory tendency.
7 Environmental variability obscures the density relationship, unless allowance is made for the feedback delay.

Recommended supporting reading

The book on "complex population dynamics" by Turchin (2003) places special emphasis on consumer–resource interactions and the cyclic dynamics frequently generated, and outlines various models that capture these patterns. Crawley (1983) focuses especially on animal–plant interactions in all of their facets, while Crawley (1992) addresses enemy–victim interactions covering predators, parasites, and diseases. A detailed analysis of the links between food resources and dynamics of herbivore populations is provided in Owen-Smith (2002b).

Programs on the accompanying CD

HVCAUGHLEY

Exercises

1 Alter the parameters for the herbivore–vegetation model outlined in this chapter to represent a carnivore population preying on a herbivore population.
2 Search the literature to obtain empirical values for the intake response parameters for a large herbivore, or any other consumer. How stable is the model incorporating these values?
3 Investigate the consequences of alternative functional forms for vegetation growth besides the logistic equation.
4 Set up a model where the amount of vegetation available for consumption is substantially less than the total amount of vegetation generating the vegetation growth dynamics. Establish how this changes the output dynamics.

6

Simulation models

Assessing understanding

Chapter features

Topics

Conceptual model; composite hypotheses; statistical relations; survival functions; environmental variability; testing the model; extrapolation; validation; identifying gaps in knowledge; simplification; omissions

6.1 Introduction

In simulation modeling, we incorporate our understanding and knowledge of some system into a fairly detailed computer representation, and observe how closely the output of this model matches the dynamical behavior of the real-world system that is being represented. This tests how good our understanding is, and also identifies gaps in our knowledge that need to be filled. If you felt that the models developed in the preceding chapters were oversimplified, you now have the opportunity to incorporate all of the relevant biological knowledge at your disposal, to bring the modeled representation closer to reality. However, some judgment is needed as to just what to include. If the information is unreliable, so also will be the model output. It helps to identify how sensitive the model is to the values of particular parameters. If changing a value makes little difference, we do not need to know the setting of that parameter too precisely. If altering values, or the form of particular functional relationships, makes a big difference, your attention is drawn to the aspects needing further study in order to make the model projections sufficiently reliable.

Putting together current information and understanding into a simulation model would be an invaluable exercise at the start of any research project, whether for a PhD or any other purpose. Conventionally such endeavors start with a literature review, establishing what is currently known about the chosen topic. This reveals apparent gaps in knowledge, justifying the need for the study. But how much difference would it make if we had this information? What is the most crucial gap in understanding? Too often modeling is left to a late stage in a biological research program, and only then is it discovered that some vital information is lacking. Engineers commonly begin their studies with a model, and then refine this model based on their findings. Conservation biologists ought to adopt a similar procedure to make their recommendations more reliable.

However, ecological systems are somewhat more complex than physical ones. Consider a simple hypothesis, for example, "an increase in food leads to an increase in population abundance." There will be conditions when this hypothesis is untrue, for example, when animals have more food available than they can utilize, or when environmental factors like predation risk restrict access to food. Some food types may make little contribution to population performance, while others that are highly nutritious or available at crucial periods may make a big difference. Hence the kinds of predictive hypotheses that we make in ecology are generally contingent upon circumstances. Modeling helps us to accommodate these contingencies when projecting likely outcomes in any specific context. A model can be viewed as a composite hypothesis to be tested. Chapter 12 in Ford (2000) provides an excellent overview of the place of modeling in ecological research.

How much biological detail do we need in a simulation model? If the model is too complicated, it becomes difficult to establish why the output is not what we expected. Was the model incorrect, was something left out, or was there simply a programming error? Experienced modelers advise starting simple and adding additional detail only when this is necessary to get satisfactory output. Biologists tend to want to include all the information that they have painstakingly gained. They regard models lacking such detail as "biologically unrealistic," and have little confidence in their predictive ability. But how reliable is all of this information based on a particular study in a specific place and time? What level of detail is appropriate? The answer depends on the objectives for which the model was developed. To advance understanding, we add details and see how much difference they make. To apply the model as an aid in making management decisions, it may be preferable to keep the model simple, and hence more general in its applicability. In both situations, it is when models fail predictively that we learn most.

The modeling foundations developed in the preceding chapters incorporated different simplifications. Descriptive equations depicted all forms of density dependence, but ignored the effects of population structure. Matrix models accommodated age- or stage-class differences in survival and reproduction, but omitted density feedbacks and environmental influences. Coupled consumer–resource models incorporated explicit links with food resources, but considered the population only in terms of its aggregated biomass.

In this chapter, we will start by considering how both density feedbacks and environmental influences can be incorporated into an age-structured population model. Thereafter we will explore a specific example and the learning that resulted from deficiencies in this model. The final product will be a relatively realistic age-structured model that could be modified to represent any mammal or bird population provided sufficient information was available.

6.2   Adding density dependence to an age-structured model

In Chapter 3, only density dependence in the overall population growth rate was considered. Recognizing that the net population growth rate is simply the difference between the effective recruitment and mortality rates, the obvious way to incorporate density dependence into the structured model is to make the age-specific survival and fecundity rates depend on the population density.

The simplest functional relationship is a linear one, that is, $a = a_0 - bN$, where a is any specific vital rate and a_0 is its potential value under ideal conditions. The steepness of the decline in survival or reproduction with increasing population density N is set by the slope coefficient b. The survival rate of young animals is expected to be more sensitive to the effects of rising density on food availability than those of mature animals with greater body reserves, and hence the value of b is likely to differ among age or stage classes. Alternatively, nonlinear density dependence in these survival rates could be introduced by making the density effect a function of some power transformation of N, for example, $a = a_0 - bN^z$. A value for the power coefficient z greater than one would produce a gradual initial decline in the vital rate, dropping more steeply toward higher population densities

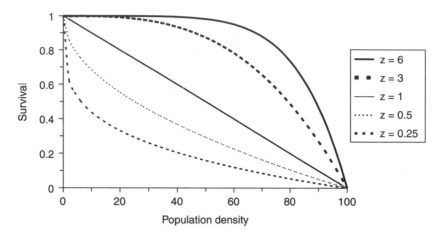

Figure 6.1   Possible forms for the relationship between age-specific survival rates and population density based on the formula $S = a - bN^z$ for different values of z.

(Fig. 6.1). This might be an appropriate pattern for mature adults. A value of less than one for z would produce a sharp initial drop flattening out toward high density levels, which might best describe the survival pattern for juveniles. Another possibility would be to introduce a threshold decline in survival or fecundity. The value of the vital rate would then remain at its maximum until some density level was exceeded, and thereafter decline linearly.

Based on these concepts, formulate a simple age-structured population model incorporating density dependence in the vital rates in a spreadsheet, following the guidelines given in Appendix 6.1. You should find that the shape of the population trend (Fig. 6.2a) as well as the overall density dependence in population growth rates (Fig. 6.2b) closely matches that described by the logistic model, if you assumed linear density dependence in age-specific survival. The only differences should be (i) some initial wiggles in the population trend as the age-class proportions in the population adjust from the arbitrary starting numbers, and (ii) a slight oscillation in the population size when it reaches the zero growth level and the age structure adjusts from a growing to a static population. Note the specific combination of juvenile and adult survival associated with zero population growth, which arise from the values of the intercept (a_0) and slope (b) of the survival functions for these age classes. Observe how alterations in these parameters change the zero growth level as well as the age-class proportions at this level (Fig. 6.2c). Growing population show a high proportion of young animals, and stable populations a greater preponderance of adults.

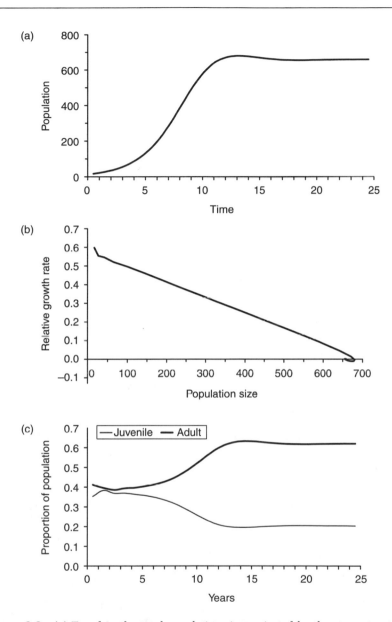

Figure 6.2 (a) Trend in the total population size projected by the stage-structured model incorporating linear density relationships for survival rates. (b) Corresponding density dependence in the population growth rate projected by the stage-structured model, after appropriate adjustments of the initial age structure. (c) Changes in the relative proportions of juveniles and adults in the population as the population trend changes from increasing to zero growth.

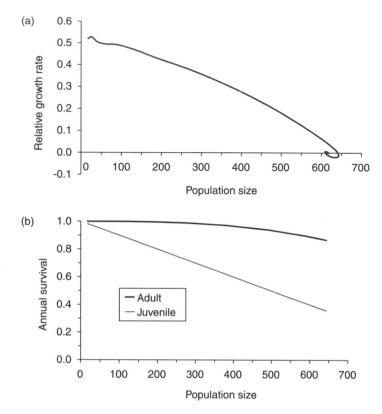

Figure 6.3 (a) Convexity in the overall density dependence in population growth rate introduced by making adult survival a convex function of density with $z = 3$. (b) Associated survival functions for adults and juveniles.

Experiment further with this model by making the survival of the adult class a convex rather than linear function of the population size. The result should be some degree of convexity in the overall density dependence in the population growth rate (Fig. 6.3). Then add a senescent class with potentially infinite survival by allowing animals within this last class to persist in it from one year to the next, but with reduced survival and lowered fecundity. You should find that this change makes a relatively small difference to the zero growth level and to the associated age-class proportions. This is because the senescent class forms only a small proportion of the total population, since few animals persist very long in this stage.

This model is now a fairly realistic representation of a large mammal or bird population. A further real-world feature still remains to be added: allowing the vital rates to vary from year to year, in response to the prevailing environmental conditions. Refer to Appendix 6.2 for guidance on how

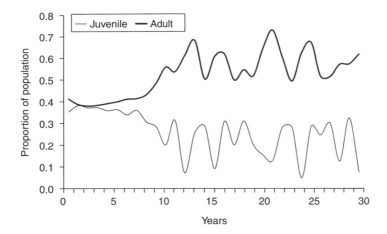

Figure 6.4 Fluctuations in the proportions of the population contributed by juveniles and adults when the class-specific vital rates respond to environmental variability.

to incorporate this adjustment into the model. The population no longer attains a stable age distribution because of the perturbations introduced by environmental variability (Fig. 6.4). Pulses of juveniles recruited during the good years move through the age structure, as also do weak cohorts born in years when most offspring died. How wildly the population size fluctuates depends on the magnitude of the environmental effect allowed in the model. If adult survival is convexly density dependent, the fluctuations become amplified because adult survival is more responsive to the variability in conditions toward higher density levels (Fig. 6.5). While a cyclic pattern in the population fluctuations may seem apparent in some trials (e.g., Fig. 6.5a), this is an illusion if the environmental variability is random. Note also how juvenile and adult survival vary in synchrony, because both are affected by the same environmental conditions (Fig. 6.6). Both stages respond relatively little to environmental variability while the population size is low.

Variability in the age structure can affect the response of the population to extreme conditions. This interaction was highlighted for Soay sheep inhabiting the St. Kilda islands (Coulson et al. 2001). Since young and old animals were most sensitive to the energy-draining effects of March gales, they incurred high mortality in years when such weather was experienced. After the die-off of these vulnerable classes, the surviving population consisted largely of prime-aged females, which are more resistant to stressful conditions. The resultant fluctuation in the proportion of the population most affected by extreme weather generated quasicyclic oscillations in abundance (Fig. 6.7).

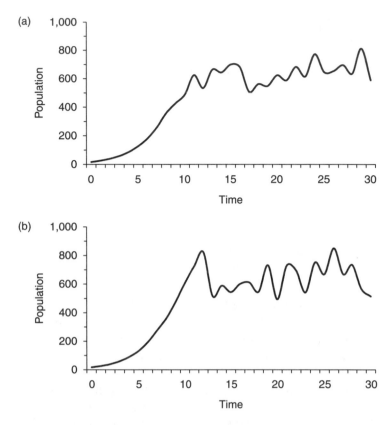

Figure 6.5   Trends in population size over time projected by models incorporating environmental variability, (a) when adult survival is a linear function of density and (b) when adult survival is a convex function of density.

Incorporating environmental influences on vital rates into an age-structured model written as a computer program is readily done through establishing two matrices. The *projection matrix* contains the vital rates that would apply under ideal conditions, that is, at low density in a benign environment. Its inherent λ projects the maximum intrinsic rate of population growth. The values for the vital rates in this matrix are then modified, depending on the population density and environmental conditions, and entered into the effective *transition matrix* applicable at that time step. Guidelines on how to do this are on the accompanying CD, and the program file AGESTR.tru provides an example. Key transitions in the model include the earliest age at which reproduction commences (the "minimum adult age"), and the age at which the mortality rate becomes elevated due to senescence.

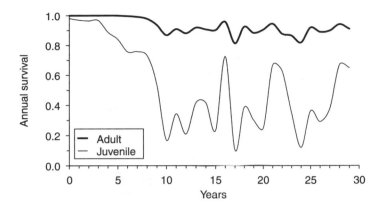

Figure 6.6   Plot of changes in annual juvenile and adult survival over time incorporating environmental influences, showing low variability initially while the population abundance remains low, and synchronous oscillations in these survival rates later causing the population to fluctuate around the zero growth level.

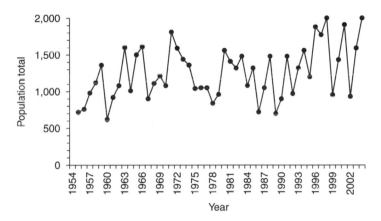

Figure 6.7   Fluctuations in the abundance of Soay sheep on Hirta Island in the St. Kilda archipelago brought about by an interaction between the population structure and extreme weather events (redrawn from Coulson et al. 2001 and Berryman et al. in press).

6.3   A specific example: the kudu model

To illustrate how such a detailed model can be used to aid understanding of the processes governing population dynamics, I will draw from my own study on a large antelope, the greater kudu (*Tragelaphus strepsiceros*; Fig. 6.8), conducted over 10 years in South Africa's Kruger National Park (KNP) (Owen-Smith 1990). Kudu are also of special interest because they

Figure 6.8 A kudu.

persist despite hunting pressure in farming areas, while not becoming "overabundant" in protected areas. This suggests that their populations are quite effectively regulated. Understanding how this works in this ungulate species could help explain apparent failures in regulation in other large herbivore populations showing outbreak dynamics or declines toward extinction.

A further special opportunity provided by kudus was the ability to recognize animals individually through variations in their stripe patterns. Additionally, females and their offspring live in closed social units typically numbering 5–15 animals. This allowed age-specific survival rates to be estimated by registering the individuals remaining alive in each group each year. The study was conducted in two study areas 100 km apart, enabling an independent replication of the population responses to changing conditions. Dispersal between female-young units was rare, so that disappearance could be equated with death for the female segment, and for young males while they remained in these units. After 2 years of age, males ranged widely in loose, shifting associations, hence their annual survival was difficult to estimate. Accordingly the basic model was restricted to the female segment.

Through good fortune, the period spanned by the study had widely variable rainfall, and the population varied more than two-fold in abundance as a result. Another serendipitous feature will be mentioned below.

Preliminary findings feature in Starfield and Bleloch's (1991) book on modeling. I will take the analysis further and induct you into how the simulation model that was constructed from the findings served heuristically to advance understanding of how various factors influence population dynamics.

6.3.1   Choosing appropriate survival functions

We need first to decide

1 which age classes to distinguish,
2 what explanatory factors affecting vital rates to include in the model,
3 how these factors are functionally related to survival and reproduction, and
4 the specific parameter values governing these relationships.

In order to assess annual variability in survival rates, kudus were grouped into the following stage classes: (i) juveniles less than 1 year old, (ii) yearlings aged between 1 and 2 years, (iii) prime adult females up to 6 years of age, and (iv) old adult females beyond 6 years of age. Initially the division between prime and old females was unknown. This was established after sufficient data had been gathered to identify the age at which the mortality rate began rising (Owen-Smith 1993b), which turned out to be somewhat younger than originally anticipated. The population registration was carried out around or shortly after the end of the dry season each year, when the animals were most readily observable. At this time the calves, born during the late wet season, were aged around 0.7 years. Hence the survival estimates refer to the following age ranges: juvenile, from conception to 0.7 years; yearling, from 0.7 to 1.7 years; prime adult, from 1.7 to 6.7 years; and old adult, from 6.7 up to the maximum recorded age of 15 years. Note that the survival estimate for juveniles, derived from the mother – offspring ratio, incorporates conception failures and fetal losses as well as mortality after birth. It assumes also that the death of a mother prior to the census automatically means the death of her calf.

Survival rates seemed likely to depend largely on food availability, governed by vegetation growth and modified by the prevailing population density. In African savannas, annual vegetation growth is controlled largely by rainfall. Hence, expressed formally, $S_a = f(R, N)$, where S_a represents the age-specific survival rate, R the rainfall measure governing production, N the effective population density, and f some unspecified functional relationship. Let us assume further that plant growth is directly and linearly related to the annual rainfall total, a relationship well established for at least the grass layer of savanna vegetation (Rutherford 1980; Deshmukh 1984). The next question is, do resources and population density influence survival rates independently and additively? Or is survival dependent more directly on the effective resource share per individual, that is, how much food is produced each year divided by the number of animals consuming it? For simplicity in the model the latter will be assumed, with one small modification: the food demand is better assessed by the population biomass density than the numerical density because calves eat less food individually than adults. Hence the effective food availability each year was indexed by the rainfall/kudu biomass ratio.

Note the similarity with the ratio K/N used to adjust the net population growth rate for the effects of increasing density in the logistic model. The carrying capacity K obviously has some relation to resource production whereas N relates to the population demand on these resources.

What about fecundity rates? It is difficult to establish the proportion of females pregnant for free-ranging animals. All that can be recorded is the proportion of females that produce surviving offspring. Kudus generally produce a single offspring annually from the age of 3 years, although 2-year-old females do sometimes give birth. As mentioned above, for the adult class failures in pregnancy, or prenatal losses of fetuses, are effectively incorporated into the measure of calf survival derived from the adult female : juvenile ratio. The uncertainty is whether to allow for some reproduction by 2-year-old females. Since during the study period only two females aged 2 years were observed with calves, and neither of these calves survived, it will be assumed that the age at first parturition is 3 years. This assumption could be relaxed when generalizing the model for other populations or conditions.

The simplest functional relationship to assume is a linear one: $S_a = a + b(R/B)$, where R represents rainfall and B the biomass density. However, a quick numerical assessment suggests that this is not the best formula to use. If $R/B = 0.5$, then $S = a + 0.5b$, whereas if $R/B = 2$, then $S = a + 2b$. This means survival rates change more with a given increment in rainfall when rainfall is high than when it is low. The opposite pattern is more likely to prevail. A log transformation of R/B makes the form of the relationship more appropriate because survival then depends on corresponding *proportional* changes in rainfall:

$$S = a + b\log(R/B). \tag{6.1}$$

However, survival cannot continue increasing linearly with resource availability because it is effectively bounded between limits of zero and one. This means that survival rates should tend asymptotically toward a value of 1.0 at very low density (or high rainfall), and a value of 0 at very high density (or low rainfall), and thus show an S-shaped (or logistic) pattern. Relationships having this form can be made linear using a logistic (or logit) transformation $\log_e\{(nS + 0.5)/(n[1 - S] + 0.5)\}$, where n is the number of animals in the sample used to estimate the survival proportion S. The number 0.5 is added because the log of zero does not exist. Hence eqn (6.1) becomes

$$\text{logit}(S) = a + b\log(R/B). \tag{6.2}$$

A third possibility is to assume that the mortality rate, that is, one minus survival, increases linearly with the inverse of the food availability per capita, that is, B/R:

$$1 - S = a + b(B/R). \tag{6.3}$$

This equation has the advantage that one can justifiably assume that the mortality rate increases without any upper limit with rising density or declining rainfall because of its time dimension. Certain death within a month is a higher mortality than certain death only after a year. In practice, the mortality rate is truncated at a value of 1.0 when populations are censused once annually.

Statistical tests indicated that these three regression relationships fit the annual estimates almost equally well, over the observed range of variation in rainfall and density, for adult and yearling survival (Fig. 6.9). However, juvenile survival seemed to reach a maximum value somewhat less than 1.0 when rainfall relative to density was very high, probably as a result of inevitable calf losses due to fertility failures, predation, and other accidents. More interestingly, survival estimates from the two study areas fell along the same regression line, because the higher rainfall in one area was counterbalanced by the higher kudu density there. As judged by the coefficient of determination R^2, variation in the annual (R/B) ratio explained over 85% of the variability in juvenile survival, and over 50% of the variability in survival of yearlings and old females, provided outlying points for 1 year were excluded. For this year (1981), survival rates of all age classes were low, despite high rainfall relative to the population density. The additional factor causing elevated mortality in the anomalous year seemed to be an extreme cold spell during the late dry season (Owen-Smith 2000). The density feedback on population growth was evident only after allowing for the effect of rainfall variability (Owen-Smith 1990, 2006).

The regression coefficients obtained for the best statistical fit to the annual survival estimates were then incorporated into an age-structured

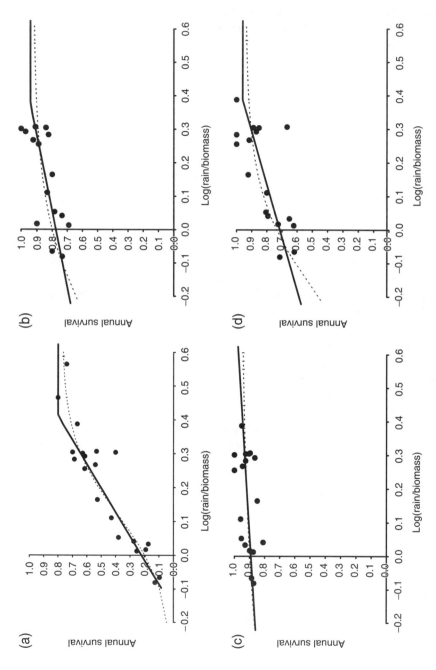

Figure 6.9 Plots of alternative survival functions for different age classes of female kudus: (a) juvenile, (b) yearling, (c) prime female, and (d) old female. Solid line indicates linear survival function with upper limit, dotted line indicates logistic survival function.

model for the chosen survival function. The population model represents 15 year classes, in line with the maximum longevity, but differentiates annual survival rates only for the four stages identified above, using the prevailing (R/B) ratio calculated from the prior density and intervening rainfall for the particular year. Equation (6.3) was adopted for the early version of the kudu model, with the survival rates bounded between upper and lower limits, because the linear relationship was easier to interpret than the logistic transformation and the latter brought no improvement in fit. Real rainfall records (annual totals over the seasonal cycle July–June) were read into the model to represent the environmental variability. To represent the total population size, the male segment was added on as a constant proportion, using the observed ratio of males to females within each age class. Subsequent analysis (Owen-Smith 2006) indicated that the mortality relationship specified by eqn (6.3) was slightly better supported and could be fit over a wider range in conditions.

The accompanying CD contains the resultant model KUDUDYN as a True BASIC program in two versions, one incorporating eqn (6.1), the other eqn (6.3). Both models require additionally a text file containing the rainfall data to be read into the model. There is a default option allowing rainfall to be held constant at the mean to enable the basic dynamics of the model to be revealed independently of rainfall variability. The output can be both graphed, and saved to a comma-delineated text file for later analysis and graphing in a spreadsheet. By making the starting numbers low (i.e., a "starting factor" <0.1) and rainfall constant, the modeled population shows slight oscillations while attaining its equilibrium density, caused by a delay in adjustment of the age structure from the proportions associated with an increasing population to those typical of a stationary population (Fig. 6.10a). Since the upper limits to the stage-specific survival rates are reached at nonzero density levels, a convexly nonlinear pattern of density dependence is manifested (Fig. 6.10b). The maximum population growth rate prevails below about one-third of the equilibrium density and amounts to about 27% per year. Note that neither the maximum growth rate r nor the carrying capacity K was specified in the model. Their effective values emerged from the specific survival and reproductive parameters and density relationships incorporated into the model.

Try replicating the output of this model in a spreadsheet, using the same functional relationships between annual survival rates and rainfall relative to population biomass density. Table 6.1 gives the fitted parameter values for the intercept and slope coefficients for each stage class. Rainfall data can be obtained from the data file KNPRAIN on the CD. Adult weights for transforming the total population N into a biomass density are 180 kg for females and 250 kg for males. You can guess at the proportional weights

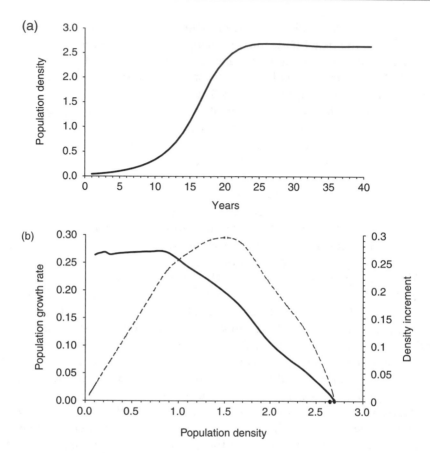

Figure 6.10 Basic output of the kudu dynamics model. (a) Population growth rate over time in a constant environment and (b) population growth rate and increment as functions of kudu density.

for yearlings and juveniles, or check the program code for KUDUDYN. You will also need to divide the population total N generated by the model by some factor to transform it into an effective population density in the region of 2–3 animals km^{-2}. From the annual rainfall (in mm) and biomass density (in kg km^{-2}) the effective (R/B) ratio each year is calculated and hence the annual effective survival rates applicable to each stage class.

This model omits possible effects of other influential factors, such as predation risk, competition pressure, and weather aspects besides rainfall. The core hypothesis inherent in the model is that kudu population dynamics are governed largely by food availability, controlled by rainfall, and modified by

Table 6.1 Specific values for the linear regression parameters for survival rates of different stage classes of kudus for the function $S = a + b \log_{10}(R/B)$

Stage	Intercept a	Slope b	Maximum value
Juvenile	0.240	1.409	0.90
Yearling	0.776	0.442	0.95
Prime adult	0.891	0.138	0.98
Old adult	0.714	0.636	0.97

Note that the effective juvenile survival rate has been adjusted to allow for a presumed constant pregnancy rate of 0.92.

density. These two factors, that is, rainfall and population density, explained much of the observed variability in stage-specific survival rates, except for prime females which showed little annual variation in survival.

6.3.2 Testing the model

How good is the model in predicting the observed dynamics of the kudu population? How do we test its validity more widely?

The first trial needed is to establish how well the model reproduces the observed changes in kudu abundance over the study period. The study population showed a low density in 1971, having just emerged from a sequence of dry years. Numbers peaked in 1978, following a series of high rainfall years. Abundance dropped during two successive drought years over 1981–3, then began recovery in 1984. The model output replicated this pattern quite closely, apart from the discrepancy in 1981, although only after the starting age structure had been adjusted for the presumed lack of calves following the 1970/1 drought (Fig. 6.11). As mentioned above, 1981 was the year when some factor besides rainfall influenced survival. Despite the discrepancy for this one year, the simulated population density almost exactly matched the observed population density in 1984 at the end of the study.

It should not be surprising that the model output corresponded so closely with the data. This merely confirms the precision with which the underlying survival functions were established. The implication is that factors besides rainfall and population density had little influence on the population dynamics of the study animals over the observation period, except in one anomalous year.

To validate the model more generally, independent data against which to compare its output are needed. This procedure is analogous to hypothesis testing in conventional science. According to strict scientific philosophy, hypotheses can never be proven, they can only be refuted. However, disproving a model is not so simple because it represents a composite hypothesis. How close a fit with reality should we expect? If the model fails the test, what portions of it should we reject? (For further discussion of these issues, see Ford 2000.)

How could we find independent data on the dynamics of a kudu population? Continuing the Kruger Park study for another decade was impractical. There were no long-term studies of kudu population dynamics done elsewhere. Moreover, comparisons between results obtained in different places could be confounded by differences in vegetation or rainfall patterns (e.g., in seasonality). The only option seemed to be to use rainfall records prior to the study to predict the low kudu density in both study areas at the commencement of the study in 1974, presumably due to the drought period of the late 1960s. The kudu densities in the study areas could be extrapolated back to 1971, using the age structure of the population at the start of the study as an indication of earlier population increments.

Figure 6.12a shows the retrodictive output, based on prior rainfall records, compared with the actual kudu density levels. The model indicated a density of 2.2 kudus km^{-2} in 1971, when in fact the projected density was only 1.3 km^{-2}. This discrepancy represents a 70% overestimate,

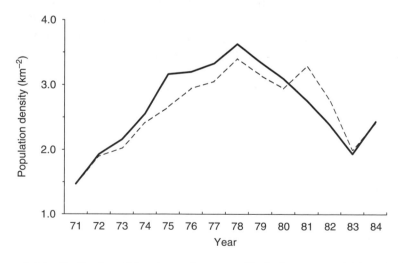

Figure 6.11   Replication of the observed pattern of kudu dynamics (solid line) by the model output (dashed line).

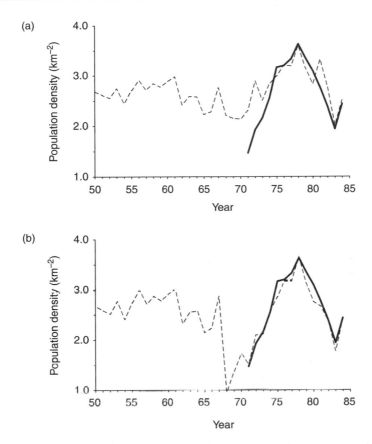

Figure 6.12 (a) Changes in kudu population density projected by the model using rainfall data prior to the start of the study (dashed line) compared with observed pattern (solid line). (b) Changes in kudu population density projected by a modified model including the effects of episodic cold weather on kudu survival in addition to rainfall (dashed line) compared with observed pattern (solid line).

which is quite a severe error. The very precise model failed its validation test rather badly. We may conclude that rainfall alone is insufficient to explain the low kudu density in 1971, unless there was some error in the model formulation.

6.3.3 Rethinking the model

The demographic reason why the simulated population density did not drop to the levels observed around 1971 is because the survival rate of prime adults was not affected much by the low rainfall. What might have caused higher mortality among prime-aged females? Possible additional factors include the following:

1 an outbreak of some disease;
2 heightened predation pressure;
3 greater competition from other species exploiting similar food resources;
4 extreme weather conditions, apart from low rainfall;
5 a degradation in the state of the vegetation, such that the rainfall was less effective.

The first "wild card" to eliminate is disease. Kudus incurred severe mortality from anthrax in the northern half of the Kruger Park around 1970, but there were no signs of disease-related mortality among kudus in the southern section where the study areas were located. It seems unlikely that an outbreak would be missed because the park authority actively suppresses these outbreaks by burning carcasses of all infected animals. It also seems unlikely that predation pressure could have been much higher prior to 1974 than during the study period, because lions were so abundant during the latter as to be held responsible for a decline in the abundance of wildebeest and zebra, their staple prey (Smuts 1978). Kudus do not suffer much competition from either giraffe or impala, the two other browsers that are common in the Kruger Park.

It is possible that the drought period of the late 1960s caused some mortality among woody plants, with a resultant depression of the food base for kudus. Although tree populations were not monitored, one would expect that the population responses of trees to improved rainfall would take several years, so that browse plants should have been recovering from the drought during the beginning of the study period. There were no signs of this.

Last, let us turn our attention to the association between low survival rates relative to rainfall in 1981 and the occurrence of an extreme cold spell in September of that year. This cold front was characterized by a maximum daily temperature that did not rise much above 14°C, coupled with rain and probably wind (Owen-Smith 2000). Perhaps similar conditions subjecting the kudus to hypothermal stress, at a time of the year when their body fat reserves were low, had occurred prior to 1971. Although such a temperature level may be comfortable for northern ungulates, kudus have to cope with maximum temperatures that frequently exceed 40°C at the start of the growing season, and hence have thin coats enabling them to continue foraging in hot weather (Owen-Smith 1998). Hence they are inadequately protected when temperature levels drop some 25°C lower. Kudus are noto-riously susceptible to mortality during cold spells around the end of the dry season.

Kruger Park's weather records showed five extreme cold spells, associ-ated with maximum daily temperatures under 16°C plus rain and wind, in the relevant period: during November in 1968 and 1969, during October in 1971 and 1973, and in September 1974. The last event occurred at the start of the study period but had no detectable effect on kudu sur-vival. Should we accept the hypothermia hypothesis, despite the vague and inconsistent evidence?

The 1974 cold spell occurred in a year when rainfall was very high and kudu density was low. Presumably animals retained abundant fat reserves, which allowed them to withstand the cold spell. The cold spells during 1968–73 occurred in dry years when the animals' body reserves had prob-ably been severely depleted by the late dry season. These conditions may have been responsible for circumstances in which even prime adult females died from cold. Hence an interactive effect of temperature and rainfall is suggested.

This hypothesis seems reasonable and is supported by the observations of relatively low survival rates in 1981, plus anecdotal reports of kudu die-offs in 1968. We cannot establish its validity statistically because there are no replicates of the 1981 event with estimates of kudu survival rates. To establish its validity would need monitoring of the kudu population to continue, in order to observe its response when further cold spells occurred. To get five repetitions would seem to require a study spanning three to five decades, judging by the single occurrence during the 1974–84 decade.

This last lucky serendipity in the kudu study is now revealed. If several cold spells had occurred during the study period, it is doubtful whether the effects of rainfall could have been separated statistically from those of cold weather and density. Since the influence of rainfall was established so precisely, it was possible to identify when additional factors affected survival rates and hence population dynamics.

6.3.4   Improving the model

A dilemma is whether to include temperature as an additional factor in the model, even though its influence cannot be confirmed statistically. To decide this, we should take into account possible purposes for which the model might be applied, recognizing the uncertainty regarding the cold spell effect.

Assume that a kudu population is being hunted or harvested for venison. If the effects of extreme cold weather are omitted from the model, and cold does indeed affect kudu survival, then every time a cold spell occurs, the harvest quota will be excessive. If the cold weather effect is included in the model, but the kudus fail to respond as assumed, the population will be underharvested that year. The consequences of the second error seem less severe than those of the first. The model that would be associated with "minimum regret," comparing these alternative scenarios, would seem to be one including the hypothesized effect of cold weather.

How do we incorporate the cold spell effect? We have reason to believe that the consequences for kudu survival rates depend not only on (i) how low temperatures drop, but also on (ii) whether the cold is coupled with rain and wind, (iii) how low the preceding rain has been, and (iv) how late in the dry season the cold spell is. We can at best make an "educated" guess and check our expectations against observations when further extreme cold events occur. Initially our guesstimates can be included in a modified model, and predictions from this model can be compared with the observed kudu density in 1971, to check for consistency. With suitable adjustments, the match with the observed kudu dynamics seemed very close (Fig. 6.12b).

However, the model including the cold spell effect failed to reproduce the observed trend in the overall kudu population in the park, from annual aerial censuses, after the study period (Owen-Smith 2000). This implicated some additional factor omitted from the model as being responsible for reduced survival among adult kudus, and elevated predation became a contender (Owen-Smith, Mason, and Ogutu 2005).

The model neglected any interaction between the kudu population and its food base, contrary to the interactive model for herbivore–vegetation systems outlined in Chapter 5. The assumption is that vegetation in the Kruger Park has been affected by kudus and other browsers for many decades, and so is unlikely to change much in response to a change in the kudu population, at least not within 1–2 decades. The outcome might differ if kudus were introduced into a new area where the plants had not previously experienced much browsing.

6.4   Simplification for management

The kudu model incorporated much information about stage-specific survival rates. How necessary is the demographic detail? The basic conceptual feature is the dependence of survival rates on resource availability governed by rainfall relative to density. A logistic model could incorporate this functional dependency if K was transformed into a variable dependent on rainfall, while aggregating the kudu population into a single total N. The

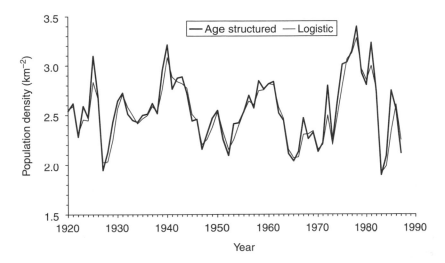

Figure 6.13   Model predictions of changes in kudu abundance over a 70-year period driven by rainfall variability using records from KNP. Output of the age-class structured model is compared with that of a counterpart logistic model with K carrying capacity dependent on rainfall.

effective values for the mean K and maximum r can be extracted from the demographic model and used as parameters for the logistic model. While the linear density dependence assumed in this model may exaggerate the maximum growth rate toward low density, if the form of density dependence is actually somewhat convex, this is inconsequential if the kudu population does not reach low enough densities for the discrepancy to become apparent. The output of this simplified model deviated surprisingly little from that of the detailed demographic model (Fig. 6.13). The small differences in the simulated population trajectory arose merely because the age-structured model generated more rapid recovery following periods of low rainfall than did the logistic model lacking demographic detail.

This deviation in the transient dynamics could be rectified using a model representing just three recruitment stages, juveniles, yearlings, and subadults, with prime and old females lumped into one adult class. The adult survival function for this model would need to be adjusted to refer to the total adult segment, with the effects of senescence on adult survival obscured. Aspects of the dynamics lost are the greater susceptibility of old adults to mortality during extreme events, and the recovery potential of the prime adults predominating after such events. Information to differentiate the survival rates of prime and old animals is lacking anyway for most populations (see Heppell et al. 2000). The output of this demographically simplified model would be almost indistinguishable from that of the age-structured model representing all 15 year classes, so as to be adequate for most management purposes.

6.5   Generalizing the model for other species

The kudu model can be easily adapted to represent any fairly similar species. All that needs to be done is to adjust the age at which the various transitions between stage classes occur, and set appropriate functional forms and parameters for the survival relationships. On the CD is a generic program for a large mammal named UNGULDYN. The statements controlling these adjustments can be found near the top of the program code, so that they can be changed easily. Almost everything lower down in the program is merely "book-keeping." If you have set the model up in a spreadsheet instead, shift the multiple columns showing the numbers in all of the age classes to one side, or to a second worksheet. This allows the focus of attention in the section of the worksheet displayed to be on the changes in total population size or biomass in response to the changing density and environmental conditions. The male segment of the population could also be represented on an additional sheet, if its independent dynamics is also of interest, for example, for trophy quotas.

6.6   Overview

This chapter should have provided you with the conceptual foundation and modeling tools to formulate a demographically structured population model that also takes into account environmental variability, for any large mammal or bird species. The model is deterministic as far as the response of the population is concerned, although the environmental variation influencing its dynamics brings in a random component. For a simplified version of this model incorporating demographic stochasticity as well as environmental uncertainty, turn to Chapter 8. An applied model incorporating the herbivore–vegetation interaction will be developed in Chapter 11.

From the material covered in this chapter, you should have learned the following:

1 Density dependence can be incorporated easily into an age-structured model.
2 Simulation models can take into account as much biological detail as is available.

3 Some simplification is always necessary, for example, kudus were lumped into four classes for assessing the effects of rainfall on survival, while the dynamics of the male segment were omitted.

4 More fundamental assumptions were the exclusion of vegetation impacts as well as interactions with predators and competitors.

5 A density feedback on vital rates was included despite weak statistical support on theoretical grounds.

6 Replicating the observed data was a test of the precision of the model formulation, but not of its wider validity.

7 Even a very detailed simulation model can fail predictively.

8 Learning commences when the model fails because omissions or discrepancies in the model formulation then become apparent.

9 Additional influences such as the cold spell effect could be included for strategic reasons even if not supported statistically.

10 A simpler model can yield predictions almost as good as those from a detailed model, provided the most salient influences are represented.

Recommended supporting reading

Models similarly incorporating both demographic structure and environmental influences have been formulated for other ungulate populations. A simple spreadsheet model for a mountain reedbuck population was outlined by Norton (1994), while other models for African ungulates incorporating rainfall influences are described by Mduma et al. (1998; wildebeest) and Georgiadis, Hack, and Turpin (2003; zebra). For models developed for northern ungulate populations, see Walters, Hilborn, and Peterson (1975; caribou), Euler and Morris (1984; white-tailed deer); Coughenour and Singer (1996; elk); Coulson et al. (2004; red deer); and Clutton-Brock et al. (2004; Soay sheep; see also Coulson et al. 2001). Among other organisms, Dennis and Taper (2000) modeled the joint effects of density and rainfall on the abundance of foxes, while Noon and Sauer (1992) reviewed models applicable to birds.

Programs on the accompanying CD

AGESTR; KUDUDYN; KNPRAIN; KUDUMRB; UNGULDYN

Exercises

1 An appendix to Chapter 1 gives a checklist protocol recommended when reporting findings from modeling exercises. To what extent does the

information that you were given above on the kudu model follow this protocol, and what is missing? Is there some additional information that you would like to have, which has not been supplied?

2 Derive a simplified model that replicates the output of the detailed kudu model adequately for management applications, either in a program or in a spreadsheet. Note that the model is defined by the regression relations between survival (or mortality) rates and rainfall relative to density.

3 Modify the kudu model to incorporate the effects of episodic cold weather, or density-dependent predation, or disease outbreaks, or any other factor that has been left out.

4 Change the formulation of the survival and/or fecundity functions used in the kudu model and observe how much difference it makes to the model output.

5 Adapt the kudu model to represent another species for which you can find the relevant information needed.

7

Harvesting models

Adaptive management

Chapter features

Topics

Surplus production; maximum sustained yield; stock reduction; dynamic pool; fixed versus variable harvests; size selection; overharvesting; safety margins; uncertainty; information costs; active adaptive management

7.1   Introduction

Populations of many animal species are harvested to supply material benefits to humans. This includes fish from the sea and waterfowl, grouse, deer, and other "game" for recreational hunting. The challenge is to provide a continuing yield of the product while ensuring the persistence of the resource. Much science and many models have been applied to aid such enterprises, yet in the fisheries context the record is one of repeated failures to sustain the resource (Roberts 1997; Fig. 7.1). In this chapter you will be inducted into some of the challenges in applying these models in

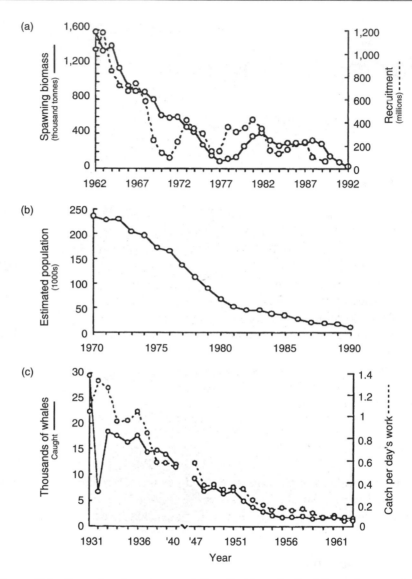

Figure 7.1   Examples of marine populations that have declined through overexploitation (from Reynolds and Jennings 2000). (a) Northern cod, (b) Bluefin tuna, and (c) Antarctic blue whales.

practice and some of the adaptive strategies advocated in responding to these difficulties.

The obstacles to be surmounted arise basically from various sources of uncertainty: (i) variability in the environmental conditions influencing population change, (ii) inadequate understanding of the processes governing survival and reproduction, (iii) imprecise information on the abundance of the resource, and (iv) weak controls over the removals that actually take place (Nichols, Johnson, and Williams 1995). Ludwig, Hilborn, and Walters (1993) outlined how scientific uncertainty, coupled with economic and social pressures, has led to inevitable overharvesting of many fish populations. These circumstances are especially acute for fisheries due to loose institutional controls over what is effectively an open access resource, that is, no specific owner bears the full costs of overexploitation. Furthermore, information on the resource abundance and dynamics is more difficult to obtain than for terrestrial populations. In this chapter you will begin by applying harvesting theory to a waterfowl population, before moving on to fisheries challenges. But first some basic principles.

7.2   Principles of "maximum sustained yield"

The exploitation of biological resources rests on the concept of *surplus production* (Hilborn, Walters, and Ludwig 1995). All populations have the reproductive potential to increase under favorable conditions. This upward growth is eventually checked by various constraints, most basically competition for limiting resources. Reducing the population distributes food and other resources more generously, enabling improved survival and greater reproductive success. These processes underlie the concept of density dependence in population growth, introduced in Chapter 3, although density-related changes in predation may also contribute. The excess of recruitment over mortality can be removed from the population without affecting its persistence adversely, provided the amount does not exceed the replacement potential. The additional mortality imposed by catches or hunter kills may be quite small if there is consequently less mortality from other causes. The improvement in survival and reproductive output *compensates* to some extent for the losses imposed by the harvest.

The simplest density-dependent model is the logistic growth equation, which generates a linear decline in the relative growth rate as the population size or its effective density rises. However, our focus of attention now shifts to the recruitment curve, that is, the number of animals by which the population increases in each time step, which indicates the surplus production available for harvesting. This population increment is the product of the proportional growth rate and the population size. The annual increase described by the logistic equation in discrete time, that is, $\Delta N / \Delta t = r_0 N (1 - N/K)$, takes the form of a parabola with a zenith when the population size is half of the zero growth level K (Fig. 7.2a). At this

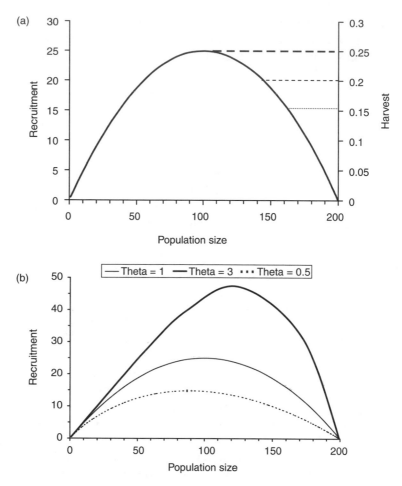

Figure 7.2   (a) Recruitment curve from logistic model with $r_0 = 0.5$ and $K = 200$, indicating three theoretically sustainable harvesting proportions, with the maximum yield equal to 0.25 as a proportion and 25 numerically; (b) recruitment curves from the theta-logistic model showing the effect of changing the value for theta.

point the relative growth rate is half of its maximum at low density r_0. Hence the maximum numerical rate of increase is $r_0/2 \times K/2 = r_0K/4$.

The traditional objective for fisheries management and similar resource cropping operations is that removals should be as large as possible but nevertheless sustainable in the long term, that is, the catch should be the maximum sustainable yield (MSY) (Punt and Smith 2001). This is commonly equated with the peak in the logistic recruitment curve. Harvest levels below this maximum recruitment are potentially sustainable, but less than what is theoretically possible (Fig. 7.2a).

To understand how this works, set up the logistic equation, or an alternative density-dependent equation, in a spreadsheet, following the guidance given in Appendix 7.1. Calculate the theoretical MSY from the above formula for the specific values of r and K that you chose. You should find that any harvest reduces the population below the zero growth level K that would otherwise be attained. If the number removed each year exceeds the MSY, the population declines progressively toward extinction. There is "knife-edge" sensitivity around the MSY level. If the initial abundance is below $K/2$, some harvest levels that could be sustained at higher population levels can no longer be supported.

In reality the shape of the density-dependent relationship could deviate from the linear relationship projected by the logistic model. This can be explored by adding a power coefficient to produce nonlinearity in the density dependence, using either the theta-logistic equation or generalized Beverton–Holt equation (see Section 3.6). For convex density dependence (θ or $b > 1$), the peak in the recruitment curve is shifted toward higher density levels, and the maximum sustainable harvest exceeds that projected by the logistic equation for the same values of r_0 and K (Fig. 7.2b). For concave density dependence (θ or $b < 1$), the peak recruitment occurs at lower density and the potential harvest is considerably reduced.

The practical consequences are as follows. For the convex pattern, a harvest quota estimated from the values of r_0 and K assuming linear density dependence would be safe because it underestimates the actual maximum harvest that could be taken. However, if the trend in recruitment when density is higher were to be extrapolated toward $K/2$, expecting a peak at this level, the actual safe quota would be overestimated. For concave density dependence, a harvest based on the observed K and estimated r_0 would greatly exceed the sustainable level. Hence structural uncertainty about the form of density dependence in either direction could lead to overharvesting. Further structural problems with fitted equations come from unreliable estimates of the effective values for r_0 and K. Overestimation of either of these parameters exaggerates the sustainable harvest level.

These principles of surplus production seem very clear when exemplified in an idealized constant environment. In the real world, the population growth rate achieved and maximum population level that could be supported fluctuate from year to year in response to environmental variability. How can harvests that are both high and sustainable in the long term be achieved in such circumstances? To explore this, we will turn to a specific example based on a waterfowl population.

7.3   Surplus production model accommodating environmental variability

The hunting of ducks and geese has developed into a huge industry in North America and is also widespread in other parts of the world. Nevertheless, much uncertainty still remains about the setting of harvest quotas (Nichols et al. 1995; Williams, Johnson, and Wilkins 1996). Some of this arises from establishing the extent to which losses through hunting are compensated through reductions in other forms of mortality. There is also relatively little understanding of how the size of the population influences the survival and reproduction of birds in nesting areas, and also later on wintering grounds. Variability in annual precipitation influences the population by affecting the number of ponds available as breeding habitat in the landscape.

Almost all of the major waterfowl species in North America incur substantial hunting impacts, so that it is difficult to establish their intrinsic population dynamics. Accordingly, for the waterfowl model we draw on a study of magpie geese in Australia's Northern Territory (Bayliss 1989). Aerial counts of this nonmigratory species in their floodplain breeding habitat were conducted in four successive years, spanning five regions. These surveys provided estimates of the annual population change, which could be related to the size of the population and to rainfall over the preceding season. The relative population growth rate was strongly influenced by the antecedent breeding density of the geese (Fig. 3.14a), as well as being affected by rainfall (Fig. 7.3). At the time of the study there had evidently been little hunting of this population.

The regression relationships relating the annual population growth both to density and to rainfall form the basis for the spreadsheet model outlined in Appendix 7.2. If you start with the simulated population at low density, its growth toward its maximum abundance level will be less smooth than described by the logistic equation in a constant environment (Fig. 7.4). Nevertheless, the amplitude of the fluctuations in abundance correspond approximately with those shown by the local goose populations. Having established basic MSY level for a constant environment, from the estimates of r_0 and K taken from Fig. 7.3a, assess how much reduction in annual harvest below this theoretical ceiling is necessary to ensure the long-term persistence of the population over a reasonably long time horizon, allowing for the effect of randomly varying rainfall. Bayliss (1989) suggested that the harvest rate needed to be reduced by 25% below the putative maximum. You should find that no precise answer can be given because of inherent uncertainty associated with the rainfall pattern. In some trials the population will persist whereas in others it will crash for the same harvest level (Fig. 7.5).

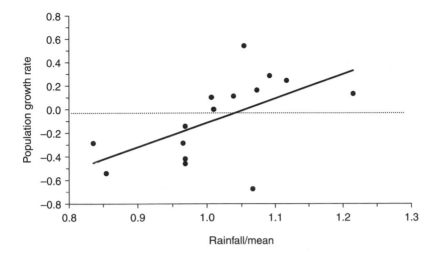

Figure 7.3 Dependence of the relative population growth rate of magpie geese on the preceding annual rainfall relative to the mean (redrawn from Bayliss 1989).

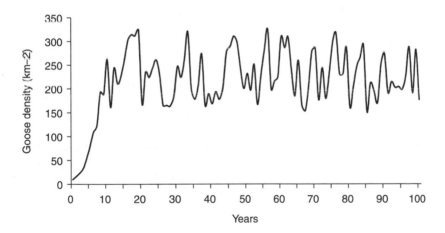

Figure 7.4 Growth of simulated goose population in an environment with variable rainfall.

Hence deciding on the optimal sustained harvest is a matter of balancing the risk of a population crash against the benefit from increased harvests if the population persists. Applying a *fixed quota* for the harvest each year carries the highest risk of overharvesting in a bad year, unless the quota is set very low, but then the benefits of the higher removals possible in good years are lost. Harvesting a *fixed proportion* of the estimated population reduces the risk because the offtake is lowered in years when the population

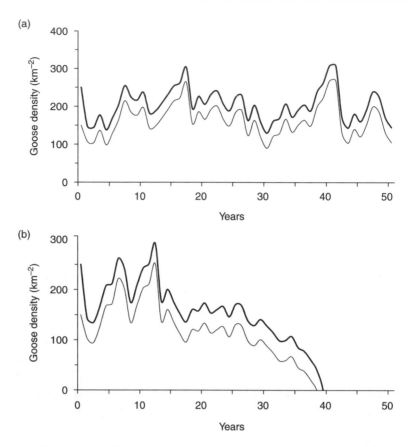

Figure 7.5 Outcome of two specific trials harvesting the simulated goose population at a fixed quota of 40 geese per year, showing trends in both preharvest and postharvest abundance. In (a) the population persisted, while in (b) it crashed to extinction, despite the identical harvest quota.

has declined. However, to do this effectively requires a current estimate of the population size. Alternatively, one could anticipate that the population is likely to be reduced in years with low rainfall prior to the breeding season, and lower the quota accordingly.

This model entailed applying an overall harvest quota to the entire population, recognizing that young geese newly recruited to the population are not distinguishable from older adults to a hunter. For other animals, additional decisions can be made regarding which segment of the population to harvest. This requires information on the population structure in terms of age and sex classes in addition to the overall abundance. This breakdown can be obtained readily for fish populations, to which we turn next.

7.4   Stock-recruitment model

Stock-recruitment models apply not only to fish populations, but also to any species for which only a portion of the population is harvested, specifically a particular sex or age class. For many ungulates, the hunting quota is restricted to males above a certain age. For fish, the harvest is restricted to individuals above a certain size by regulating the mesh size of the fishing nets used. This allows smaller and hence younger fish to escape being harvested until they exceed the minimum size. Size limits are also commonly imposed on the catches of other organisms, for example, rock lobsters.

In this approach, the harvestable segment is conceptualized as constituting the *stock*, and the rest of the population is viewed as providing recruitment into this *dynamic pool*. Furthermore, the value of the stock can increase as a result of growth in size, as well as through ingrowth from smaller classes. Removals through catches as well as natural mortality reduce the size of the stock, measured in biomass units rather than numerically. In simple mathematical outline,

$$B_2 - B_1 = (R + G) - (M + C), \tag{7.1}$$

where B_2 is the harvestable stock biomass this year; B_1, the harvestable stock biomass left the previous year; R, the recruitment to the stock; G, the growth in weight of the stock; M, the loss due to natural mortality; and C, the catch removed from the population.

The stock-recruitment function R is generally described by some form of density-dependent relationship. Indeed, the Ricker and Beverton–Holt equations introduced in Chapter 2 were derived to represent such relationships for fish populations. Both the Ricker (or exponential logistic) and Beverton–Holt (or hyperbolic) equations produce humped recruitment curves, but with production declining more sharply above the zero growth level with the Ricker formulation (Fig. 7.6). Accordingly, the Ricker equation has a higher propensity to generate oscillatory dynamics, which might be appropriate for fish species laying hundreds or thousands of eggs annually per adult. The Beverton–Holt equation is regarded as appropriate for long-lived fishes with relatively low recruitment rates. While the simple hyperbolic formulation does not produce oscillations, this can be altered through the incorporation of a power coefficient, as described in Chapter 3. A further mechanism promoting oscillations in abundance will be identified below.

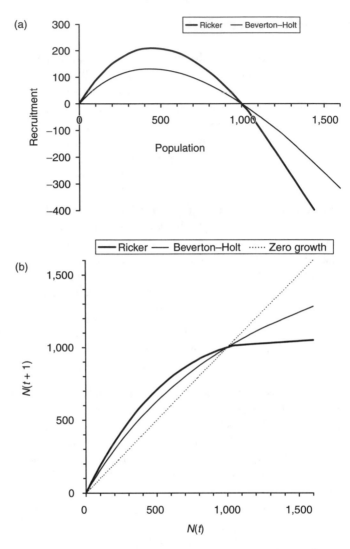

Figure 7.6 Comparative recruitment curves from the Ricker and Beverton–Holt equations for identical values of r (0.7) and K (1,000), plotted (a) as recruitment versus stock size and (b) as projected versus current population.

Detailed models represent sequential stages of recruitment from eggs through fry to mature adults (Fig. 7.7). They require estimates of growth, survival, and fecundity rates for each stage. Fisheries biologists believe that the density feedback arises mostly through influences on early survival during the posthatching period. Cannibalism from adults of the same species can make a major contribution, together with predation from other

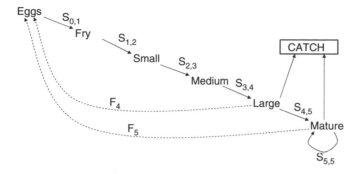

Figure 7.7   Outline of successive recruitment stages for a stock-recruitment model of a fish population. S represents survival rates and F fecundity.

species. Subsequent survival, growth, and reproduction seem to be largely independent of the population abundance, because the adults are widely dispersed through the ocean. The survival of young fry is also sensitive to water temperature, the flow of ocean currents, and other environmental factors, which can cause wide fluctuations in recruitment success from year to year.

Ocean fish are obviously difficult to census directly, the exception being salmon which can be counted when they swim up rivers to spawn. Hence population abundance is usually estimated indirectly from the rate at which fish are caught during harvesting, standardized as the "catch per unit effort" (CPUE). The effort is generally indexed as trawler days spent fishing. This is obviously a crude measure, subject to variability in harvesting efficiency as well as stock size. If the population is highly aggregated, and searching is directed toward large schools, the CPUE may show little change until the resource is substantially depleted.

The form of the recruitment function is generally established either from (i) the decline in catch rate with increasing fishing effort (Fig. 7.8a), or (ii) from the proportion constituted by the youngest cohort among the fish caught. However, these recruits were hatched as fry several years earlier. Hence an additional sample of the population structure is needed encompassing smaller size classes, so that recruitment can be related directly to the population size generating it. Wide annual variability in recruitment at the fry stage results in rather vaguely fitted functions (Fig. 7.8b). For many fish populations, the stock is constituted largely by a few cohorts produced in favorable years. These cohorts undergo steady attrition through harvesting coupled with natural mortality. If the next recruitment episode is delayed, the catchable stock may become severely depleted causing the yield to drop.

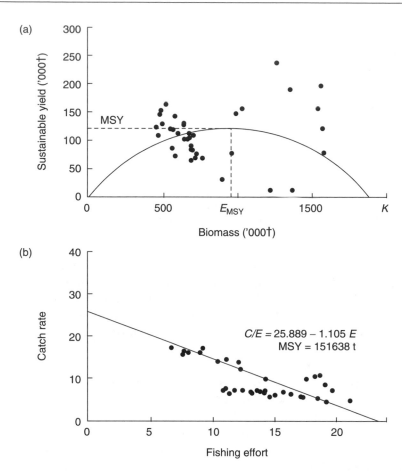

Figure 7.8   Models fitted to data for hake population off west coast of South Africa to estimate the MSY. (a) Identification of the MSY in relation to the corresponding stock biomass with fitted recruitment curve. B_{MSY} is the stock level providing the MSY; (b) catch rate (annual catch divided by the corresponding fishing effort) regressed against the annual fishing effort (trawler days) giving the MSY as $a^2/4b$, where a is the catch rate at the Y-axis intercept and b is the slope of the regression with increasing effort (from Punt and Smith 2001).

7.5   Policies for setting the harvest quota

Fishing quotas may be set in three basic ways. A *fixed quota* may be partitioned among the number of boats allowed to fish. This requires reliable reporting of the catches obtained by different operators. Whale catches were generally managed in this way. Alternatively, the fishing *effort* may be restricted by licensing the number of days each vessel is allowed to fish (equivalent to restrictions placed on the length of the hunting season for game mammals and birds). This policy is aimed at cropping a constant proportion of the stock, so that the catch drops when the stock declines, facilitating recovery. The assumption that the harvest rate is directly proportional to fish abundance breaks down if the stock is aggregated and searching is directed toward the aggregations, such as large shoals. Policies aimed at proportional harvesting are less popular than fixed quotas with the industry because of variability in economic returns between years.

The safest policy for sustaining the resource is to set a *constant escapement*. This entails suspending harvests once the stock falls below some specified level, generally close to the stock size yielding the projected MSY. The quota is determined by the extent to which the stock exceeds this level. Theoretically, this policy gives the greatest long-term yield from any fluctuating resource (Clark 1990; Lande, Saether, and Engen 1997; Ludwig 1998). In practical application, it creates difficulties for the industry because the allowable harvest varies widely from one year to the next. Boat capacity restricts catches during the years when high harvests are allowed, while jobs are in jeopardy when the quota is curtailed. Practical difficulties also arise in establishing whether the resource has fallen below the critical level.

7.6 Adaptive management responses

As for the goose model, practical problems arise in applying models when annual recruitment fluctuates, with consequent variation in the size of the stock. The challenge is especially acute for fisheries, due to unreliable estimates of the stock and of the parameters governing recruitment. What policy should be adopted for setting quotas, in the face of such uncertainty?

One possible response is to start with an approximation of the sustainable harvest, and refine this from experience gained over time. This can be labeled *reactive* adaptive management. *Proactive* management entails acquiring the information needed to set the quotas in advance of, or in parallel with, the operation, via supporting scientific research. Of course, the initial scientific estimates may be incorrect, hence continuing research is needed, especially to establish the effects of environmental variability. However, research is costly, and the economic returns of research investments may be questioned.

A third approach, advocated initially by Holling (1978), is termed *active* adaptive management. This entails carrying out manipulations of the population as a component of management. Restated by Hilborn and Walters

(1992), one cannot determine the MSY from a fish stock without over-exploiting it beyond the bounds of sustainability, to reveal where the threshold for diminishing returns lies. A cost will be incurred in terms of lost production, until the resource recovers. Nevertheless, this may be substantially less than research costs because it can be offset by the economic gain from the initial overexploitation. This form of experimental management should be carried on only a portion of the stock, if discrete subpopulations can be identified.

All information, whether acquired through scientific research or active adaptive management, has some degree of uncertainty associated with it. Estimates of annual survival or recruitment are subject to sampling error, as well as variability between years. Population abundance can only be indexed crudely, through the CPUE. Years with extreme conditions are experienced infrequently, and hence are difficult to replicate. If active reduction of the population toward an unsustainable level happens to coincide with environmentally induced failure of recruitment, the population recovery may be long delayed, or even fail to materialize.

7.7   Stock-recruitment models for fish populations

7.7.1   Simple structured harvesting model

The stage has come for you to address these challenges with the aid of a structured population model. Start with a very simple model representing three life history stages – newly hatched fry, small fish, and mature fish – plus the eggs from which the fry hatch, following the outline in Appendix 7.3. Ignore the possibility that some fish could remain small a year later because of restricted growth. Hence all of the fish reach reproductive maturity 2 years after hatching, thus exhibiting a fast life history. Accordingly, the reproductive potential should be set quite high, and survival relatively low. Allow only survival into the fry stage to be influenced by a density feedback, presumably related to cannibalism by spawning adults on the fry after hatching. The survival of the fry depends additionally on environmental conditions. The high fecundity based on the number of eggs laid per female needs to be countered by a low survival in the early fry stage; otherwise the inherent population growth rate is so high that chaotic dynamics are generated. Note that the oscillatory tendency is enhanced by the 2-year lag between when the fry are produced and the density feedback after these fry reach the reproductive stage.

Establish first the MSY that could be sustained in a constant environment, as determined by the parameter values and density function that

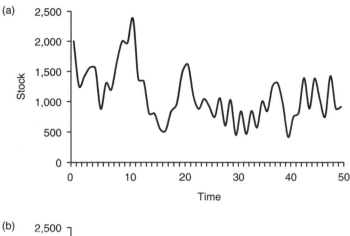

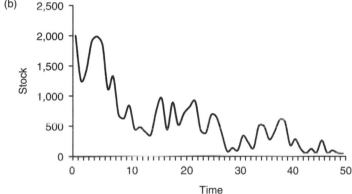

Figure 7.9 Model output from two trials harvesting a simplified fish population at the same fixed quota in a variable environment. (a) Stock is sustained, (b) stock crashes.

you assumed. You should find that virtually the entire adult stock can be harvested annually because it can be replaced by small fish growing into the adult stage and reproducing before being caught. Eventually a higher quota brings no improvement in the mean annual harvest, when the entire stock of mature fish is being removed each year. Then allow environmental variability, affecting fry survival, and observe how this alters the mean annual yield achieved by different quotas. You should find that higher quotas beyond some level lead to a reduced yield, once the stock gets held at a low level from which it cannot recover. However, whether any particular harvest quota is sustainable cannot be established with certainty over any limited time period because the random environmental variability changes the outcome when trials are repeated (Fig. 7.9).

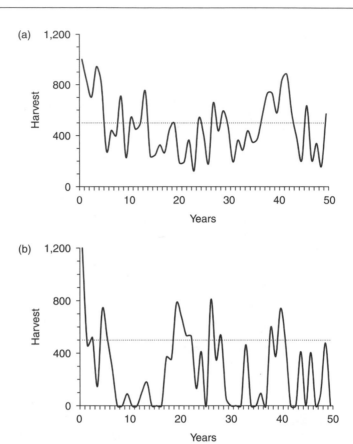

Figure 7.10 Model output showing greater variation in harvests obtained from (a) a fixed effort policy compared with (b) fixed escapement.

Thereafter, explore the benefits of assigning instead a fixed effort, in terms of the proportion of the stock removed annually. You should find that this is much less likely to lead to a crash in the population because of the reduced catch when the stock becomes low. However, the annual yield now fluctuates as widely as the stock of mature fish (Fig. 7.10a), while the composite yield achieved over any specific period also remains quite variable.

Last, explore the consequences of a constant escapement strategy. This is guaranteed to avoid a collapse in the stock and should ultimately give the highest long-term yield. However, it requires an annual estimate of the stock size and produces even wider annual variability in the harvest (Fig. 7.10b).

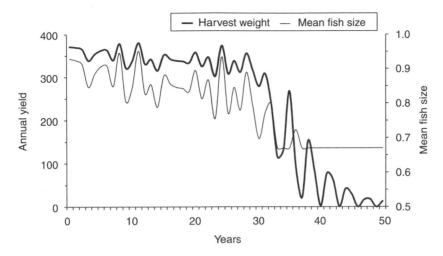

Figure 7.11 Output from dynamic pool model showing change in harvest weight and mean size of the fish harvested as the stock becomes overexploited.

7.7.2 More elaborate model with multiple recruitment stages and catch distributed over more than one stage

Longer-lived fish species may take several years to reach reproductive maturity. Hence it becomes necessary to represent these additional recruitment stages in the model. Moreover, adults may continue growing, with fecundity increasing with larger size. The relative proportions of various size classes in the catch affect the weight of the harvest, and hence the economic return. The elaborations needed to the model are described in Appendix 7.4.

Using the experience that you gained from the simple model, establish the MSY that can be obtained from this population, for various harvesting policies. An additional indicator of the state of the stock is the mean size of the fish caught, determined by the proportion of large but not fully grown fish in the catch. A reduction in size suggests that a collapse in the stock is imminent (Fig. 7.11). You can investigate further the consequences of changing the mesh size of the fishing nets, both for the harvests obtained and their sustainability.

Then set up a third model with fewer recruitment stages, higher fecundity, and lower survival to represent a fast-maturing fish like a sardine or pilchard. Experiment with all three models to establish general principles guiding the setting of maximum sustainable harvests.

The random number sequence incorporated into these models has an even distribution, meaning that extreme values are just as likely as median ones. This exaggerates the frequency of occurrence of extreme conditions. On the other hand, successive numbers are independent, whereas in the real world adverse conditions can persist for several years, amplifying the consequences. The random numbers could be replaced by some measure of environmental variability drawn from actual observations, such as rainfall for terrestrial systems, for greater realism.

The models also ignore the possibility of depensation, that is, a decline in the population growth rate toward low abundance levels (see Section 3.8). This means that extirpation could result below some threshold abundance. The evidence for such "Allee effects" in fish populations is equivocal, but the possibility cannot be ignored (Myers et al. 1995; Lierman and Hilborn 1997).

7.7.3   An adaptive management game

In the spreadsheet models that you generated, all of the parameters and functional relationships governing the output from the model were clearly visible. In the real world, scientists advising on fisheries management have to deal with much uncertainty in the information upon which they base their harvest recommendations. What adjustments are needed to cope with unreliable estimates in additional to environmental variability?

On the accompanying CD are two models written as computer programs, bound into executable format with the program code hidden. Based on the output data, you must decide what quota to apply. The basic information provided concerns the size of the harvest obtained each year. You can choose to harvest either through a fixed annual catch or through a fixed effort representing a proportional harvest. This policy can be adjusted each year, or maintained at a particular level from that year onward, over a 100-year period. You can request additional information about the population. However, all information has a cost, paid as fish subtracted from the harvest. The information that can be supplied concerns (i) the stock size, (ii) the survival rate of adult fish, (iii) the fecundity in terms of number of eggs produced per mature female, and (iv) a relative assessment of the state of the environment. However, as in the real world, the information that you receive incorporates sampling errors. You can confirm this by repeating your request for the same information, and perhaps establish how reliable the answer is from repeated samples (but at a cost). You must also decide whether it would be worthwhile to adopt an active adaptive management strategy, that is, deliberately overharvesting the population to obtain information on its recovery potential.

One of the programs has been written with parameters representing a slow-growing fish species, labeled suggestively as HARVCOD. The other program, labeled HARVSARD, represents a faster-growing fish species. These programs incorporate a threshold level of abundance, which effectively represents extirpation of the resource, that is, there is no recovery if the stock falls below this level. Apply the experience that you gained from the spreadsheet models in an attempt to secure the highest sustainable yield from these hypothetical populations. The computer program advises you of your success in terms of the total harvest obtained over the 100 years, provided the stock is not extirpated.

7.8   Overview

Despite all the limitations that you encountered, a quota based on the MSY concept remains widely applied in fisheries management, but as a limit to the total allowable catch, incorporating an appropriate safety margin and adjustments of the harvest if necessary to maintain this margin (Punt and Smith 2001). The reference points are the stock biomass associated with MSY, and the maximum fishing mortality that can be imposed at this level. Harwood and Stokes (2002) offer some further suggestions for coping with uncertainty, via more rigorous procedures for fitting models to the available data, allowing risks to be quantified.

From this chapter, you should have learned the following:

1 A harvesting strategy entails both assigning an optimal stock level from which the surplus production is generated and an annual quota obtained from this stock.
2 Real populations may show maximum recruitment at a level different from the $K/2$ suggested by the logistic model.
3 Fixed quotas that might appear sustainable in a constant environment commonly result in overharvesting when environmental variability causes recruitment to fluctuate widely from year to year.
4 Increasing the safety margin reduces the economic gains.
5 There is no analytic way of deciding on the safety margin because future environmental variability is unknown.
6 All scientific information obtained about the population incorporates some degree of uncertainty due to sampling error.
7 While either fixed effort or escapement policies provide greater security against overexploiting the resource, they are economically disruptive because of the wide annual variability in catches.
8 MSY concepts can nevertheless be applied as reference points, with appropriate safety margins to accommodate variability.

Recommended supporting reading

Many books have been written on harvesting theory and its applications. Clarke (1990) is a standard mathematical treatment emphasizing economic aspects, while Lande, Engen, and Saether (2003) review ecological theory underlying sustainable harvests in the face of environmental variability. Milner-Gulland and Mace (1998) give a more easily readable coverage of both ecological and economic principles, including case studies, while Reynolds et al. (2001) present several excellent reviews and case histories. Walters (1986) outlines principles of adaptive management with a focus on fisheries, and Hilborn and Walters (1992) offer a comprehensive coverage of all aspects of fisheries management. A detailed harvesting model for an ungulate population subject to illegal as well as legitimate removals is described by Mayaka et al. (2004).

Programs on the accompanying CD

HARVFISH; HARVCOD; HARVSARD

Exercises

1 Formulate a harvesting model for some other animal type, for example, a gamebird or a deer. Evaluate whether the optimal harvesting policy differs from that for fisheries. In particular consider the consequences of harvesting a particular population segment, for example, trophy males from a big mammal population.
2 Use the executable programs HARVCOD and HARVSARD as the basis for a management game, played between individual class members or teams. This entails applying what you learned from the spreadsheet models, over some specified time horizon, say 50 years. You have to decide how much to pay for information, whether to engage in active adaptive management, and just how much risk you are prepared to take that the population might collapse. You have just ONE opportunity to exploit each fish population, hence your strategy must be carefully planned. Of course, the outcome entails some component of luck because the random number generator determines the actual sequence of environmental conditions experienced during any particular period.
3 Contrast the net economic gains from these specific strategies applied to some hypothetical fish population over an extended time period: (i) a rigid quota, either a fixed catch or a fixed effort, (ii) reactive management

adjusting the annual harvest in response to variability in the catch obtained, (iii) active adaptive management, which entails obtaining information about the population using extreme harvests as described above, and (iv) research-based management, using scientific information obtained about the population or its environment to guide the harvest quota.

8

Population viability models

Risk analysis

Chapter features

Topics

Extinction vortices; demographic stochasticity; environmental variability; episodic disturbances; Monte Carlo simulations; persistence probability

8.1 Introduction

In the previous chapter, you confronted the uncertainty in population dynamics associated with environmental variability, affecting harvest levels. A more fundamental source of uncertainty is exposed when the population size becomes small, manifested through demographic processes. In the dodo conservation exercise of Chapter 2, you recognized that two eggs could both hatch males, despite a 1 : 1 sex ratio on average, making the population nonviable. Similarly, even though the chance of survival might be high, by chance the odds could turn out that both birds die. This meant that there was an appreciable risk that the whole conservation exercise could fail.

Hence when a population becomes small, there is a heightened risk that it will go extinct, or at least drop to low enough levels for the Allee effect to come into play (see Section 3.8). For conservation purposes, it can be important to estimate just how great this risk is. The World Conservation Union (or IUCN) has formally categorized the status of threatened species as follows (Mace and Lande 1991):

1 *Critical*   50% likelihood of extinction within 5 years, or two generations (whichever is longer);
2 *Endangered*   20% likelihood of extinction within 20 years or 10 generations (whichever is longer);
3 *Vulnerable*   10% likelihood of extinction within 100 years.

Accordingly, developing models for population viability analysis (PVA) has become a major activity in conservation biology, and the subject of several recent books. These models differ from deterministic models in that the outcome is uncertain, dependent on various chance (or *stochastic*) events. You have already accommodated unpredictable environmental variability into some of the models that you developed previously, using a random number generator, and have observed how the output changes each time the model is run. Generally four different forms of stochasticity are distinguished in population viability models:

1 *demographic stochasticity*, arising from chance variation in births, deaths, and the sex of the offspring;
2 *genetic stochasticity*, as a consequence of inbreeding, leading to chance exposure of deleterious genes;
3 *environmental variability*, unpredictably affecting survival chances and reproductive success; and
4 *catastrophes*, causing severe mortality when they occur at unpredictable intervals.

Shaffer (1981) labeled these processes "the evil quartet," while Gilpin and Soule (1986) suggested that they generate a "vortex," drawing the population down toward extinction through their synergistic effects.

The models that you will develop for this chapter emphasize the uncertainty in the outcome dependent on the chance or stochastic element inherent in the above processes, revealed especially when populations become small. We need to incorporate these stochastic elements in order to project the relative likelihood of different outcomes. Hence models need to be run repeatedly in order to estimate the probability of a particular outcome, such as population dropping to zero, from the frequency with which it occurs. These are sometimes called "Monte Carlo" trials, after the European city-state where gambling was encouraged when this was prohibited elsewhere.

Spreadsheets can accommodate stochastic variability through the random number function, as you have seen. However, they cannot readily handle the variable number of individuals constituting the population, in a situation where the survival and reproductive contribution of each individual has to be assessed using a different random number. This is much more easily handled using the flexibility of a programming language, as you will learn.

8.2   Demographic stochasticity

When the population is large, the proportion surviving is closely similar to the probability of survival. If you toss a coin 1,000 times, the number of trials yielding heads will not differ much from 500. However, if you tossed it only 10 times, the outcome could deviate quite far from 5 out of 10. Similarly, with regard to the sex of the offspring produced, the likelihood of a son or a daughter is almost exactly equal for most animals. However, if the population consisted of only 5 females, there is a 3% (0.5 raised to the power 5) chance that they will produce 5 sons, and hence leave no female descendants to maintain the population.

For survival, the odds generally differ somewhat from 50 : 50. Based on the odds, a random number generator can be used to decide whether each individual either lives or dies. All numbers between 0 and 0.999... are equally likely, hence in a very long sequence 60% of the numbers will be less than 0.6, and the remaining 40% equal to or greater than 0.6. Accordingly, to apply a survival probability of 0.6, survival occurs *if* the random number obtained is less than 0.6, and death otherwise. For each individual in the population, we need to call up a separate random number (assuming survival chances are independent). Based on the outcome, that individual either survives and remains in the population or dies and drops out.

An elementary model of demographic stochasticity can be set up easily in a spreadsheet, as outlined in Appendix 8.1. The model projects population size one time step later, from the assigned probabilities of survival and fecundity relative to the sequence of random numbers generated. The smaller you make the initial population, the greater the chance that numbers will drift toward zero, that is, population extirpation. While you can extend the projection of future population sizes for a few years, the limitations of the spreadsheet format soon become exposed. The important feature to note is how a different pattern is obtained in each successive trial even with the vital rates and starting size held constant.

The True BASIC program RANDOMWALK presents a very similar model of a "random walk" in population size. If values are selected for the survival probability and reproductive output such that the expected population growth rate is zero, population extinction is inevitable, whatever the starting population size, the only uncertainty being how long this will take. The expected population growth rate at low numbers must be positive in order for there to be a high chance for the population to increase sufficiently for the likelihood of extinction to become vanishingly small.

8.3   Environmental variability and catastrophes

Survival chances and reproductive success depend on the prevailing environmental conditions. In adverse years, the probability of survival might be lowered sufficiently for the expected population growth rate to become negative. If bad conditions persist, the population may become reduced to a level where the effects of demographic stochasticity become exposed.

In African savanna regions, rainfall controls vegetation growth, and hence the amount of food produced to support herbivore populations. Accordingly, the prevailing annual rainfall total strongly influences the population trends of many African ungulates (Mills, Biggs, and Whyte 1995; Ogutu and Owen-Smith 2003), as well as specific survival rates (Owen-Smith 1990; Owen-Smith et al. 2005). In the northern hemisphere, atmospheric conditions associated with the North Atlantic Oscillation (NAO) affect the snow cover and likelihood of extreme cold exacerbated by strong winds, and hence also the dynamics of herbivore populations (Ottersen et al. 2001). Weather conditions during the nonbreeding season have a strong influence in the survival of many bird species in the northern hemisphere, while in arid regions weather during the breeding season has a strong influence on recruitment success (Saether et al. 2004).

Superimposed on normal fluctuations in the weather conditions are extreme years when rains fail or violent winter storms develop. Across southern Africa, severe droughts are associated with "El Nino" conditions related to elevated sea surface temperatures in the eastern Pacific Ocean, occurring at irregular intervals of 5–10 years. This can result in extreme crashes by many herbivore populations, especially if animal movements are restricted (Walker et al. 1987). The same weather anomaly leads to severe floods in parts of East Africa, also with potentially disastrous impacts on many animal populations. Extreme events of this kind occurring at irregular intervals may be termed catastrophes, if their effects are sufficiently great. However, the distinction between "normal" variability and "abnormal" extremes is not clear cut. For modeling purposes, it is nevertheless useful to distinguish between episodic events, with exceptionally severe effects on survival when they occur, and the more regular pattern of environmental variability.

More obvious catastrophes are earthquakes, tsunamis, cyclones, and outbreaks of exotic diseases. The introduction of the virus causing rinderpest from Asia into Africa in the late nineteenth century eliminated over 90% of many ungulate species from Ethiopia to the Cape (Dobson 1995). Chance circumstances or coincidences in the timing of such events can greatly affect the extinction risk that they impose.

8.4   Genetic stochasticity

Genetic diversity becomes progressively lost in small populations through chance processes in the genes that happen to be passed on from one generation to the next, termed genetic drift. The extent of the loss depends on the period for which the population bottleneck persists. Genetic variability is lowered further by inbreeding, that is, the mating of individuals sharing many genes in common. The reduced heterozygosity may lower the ability of animals to resist pathogens (Coltman et al. 1999), or affect the viability of offspring (Ralls, Brugger, and Ballou 1979). The homozygosity that develops also exposes the effects of deleterious genes, should these happen to occur in the gene pool, adversely affecting survival and reproductive success. However, once thus exposed to selection, such genes can be purged from the gene pool. Populations that have survived through periods of low numbers are unlikely to retain these alleles.

The demographic consequences of reductions in genetic variability are not easily modeled. A particular population may or may not retain the bad genes. The genetic loss may be inconsequential if the population quickly recovers to higher numbers. The lowered evolutionary potential may only become apparent many generations later. Remnant populations can be managed to restrict the loss in genetic variability, for example, by selecting which individuals breed, as was done for the California condor. Recognizing these complications, genetic effects will be omitted from the models developed in this chapter.

8.5   Population viability models

8.5.1   Basic small population model

Various software packages have been produced specifically for population viability assessment, the best known being the program VORTEX originally developed by the Chicago Zoological Society (Lacy and Kreeger 1992). These prewritten programs are essentially "black boxes," with hidden design features generating the output that they produce. By the end of this chapter, you should be empowered to develop an appropriate model for the specific organism or situation that you might be studying.

The first step is to decide on the appropriate population model to use. Population structure is important because whether the population consists mainly of adults or juveniles can make a big difference to its chance of persistence. However, rarely do we have any information on longevity or the effects of senescence on survival and reproduction. Hence for an animal population, we do not need more detail than provided by an "Usher" matrix (see Section 4.2.3). Hence even for a large ungulate, we need distinguish no more than three recruitment stages plus two reproductive stages: (i) newly conceived, (ii) juvenile, (iii) immature, (iv) subadult, and (v) adult. For an annual time step, these distinctions imply that animals are born the year after their conception, remain dependent on the mother as juveniles for a year, and potentially conceive their first offspring as yearlings to first give birth when 2 years old. By 3 years of age, all individuals have matured and produce offspring annually from then on. For very large, long-lived animals like rhinos, these stages may extend over several years, as you will observe in Chapter 11. Small deer can conceive as juveniles and give birth at 1 year of age, meaning that the subadult stage can be omitted. Although most models only count animals after birth, I find it helpful to distinguish processes governing conception from those affecting survival through gestation and around the time of birth. After all, life begins at conception, even though the zygote may be hidden inside the mother or inside an egg.

The second step is to decide on how to allow for density dependence. Often there is no information on how density affects the population. Nevertheless, there must be some population size at which the expected growth rate becomes zero. This ceiling or saturation density can make a big difference to the output of the model because it limits how large the population can become if the area available to it is restricted. As you have learned, in structured population models, the density feedback is incorporated by making the specific survival rates a function of the population size relative to some upper limit. However, mortality losses may become close to zero at a population level somewhat above zero, producing a threshold pattern of density dependence (as you saw for the kudu model outlined in Chapter 6).

The third step is to decide on how environmental variability affects the population. It may affect the potential carrying capacity for the population, as rainfall does for herbivores via its effects on plant growth. Alternatively, it may affect survival chances directly, with varying effects on different life history stages. The index of the annual climatic conditions may be a direct measure like rainfall, or some index of a set of conditions like the NAO or El Nino – Southern Oscillation (ENSO), values of which can be obtained from web sites (back into the past, but obviously not for the future!). Episodic extreme events can have a superimposed effect when they occur, as we assumed for the cold spells apparently reducing the survival rates of kudus relative to rainfall (see Chapter 6).

Last, at the core of the model is demographic stochasticity. This means that instead of assigning survival as a proportion of the animals within a particular stage, we have to consider whether each individual survives or dies using a random number outcome relative to the survival probability.

The program SMALLPOP has been written to allow you to explore these various effects. The basic parameter values have been set to represent a kudu population, simplified from the more detailed model developed in Chapter 6 in terms of the age classes represented. Nevertheless, if you wish you can extend the number of adult classes represented by adjusting the "Maxage" assignment at the top of the program. However, you will find that there is little to be gained by doing so. The model automatically allows persistence in the last age class, which could be assigned a lowered survival to allow for the effect of senescence. The setting "MinAdAge" assigns the age at first reproduction. The basic projection matrix is displayed during the run, and you have the option to change the basic parameter settings for fecundity and survival at this stage. The effective fecundity setting incorporates the survival rate of adult females from one year to the next, assuming that the death of the mother automatically means the loss of the zygote. The juvenile and immature survival rates are automatically adjusted relative to the adult survival rate. Note that this model represents only the female segment.

The program offers additional options during the run, governed by your answers to questions appearing on the screen. You can switch off all stochastic influences and thereby examine the output dynamics generated by the deterministic core of the model. By additionally disabling the density dependence, you can establish the maximum growth rate for the default survival and fecundity rates (or the changed rates that you set during the run). You can set the maximum population size ("saturation biomass") as large as you like, in units of adult body mass (allowing for the differing food requirements of different age classes). By making the starting population size small relative to this maximum, and allowing density dependence, you can observe how the population grows toward carrying capacity, and the corresponding influence on the relative population growth rate, without the distracting effects of stochasticity. The default saturation biomass represents a single kudu herd of around 10 females and young. The combination of vital rates associated with the zero growth level, produced via the density-dependent functions for each stage class, is shown on the screen at the end of the run. The density feedback is set such that the strongest effect is on the survival of the juveniles from conception through birth. The trend line will probably appear a bit jagged, especially while the population size remains small, because numbers are rounded to represent whole individuals. This cause can be confirmed by raising the saturation biomass level to 1,000 or more, thereby reducing the effect of each individual added or subtracted.

You can observe how demographic stochasticity alone affects the dynamics by choosing the "Stochastic" option, while excluding environmental variability. You will find that the population size now varies erratically from year to year, despite the constant environmental conditions, with the amplitude of the fluctuations depending on the saturation biomass level (Fig. 8.1). Then, by optionally incorporating the alternative forms of environmental variability offered, you can see what additional effect each has on the output dynamics, separately or in combination. The environmental variability is governed by real rainfall records, included as data at the end of the program, and so is not genuinely stochastic – the same pattern will apply in every run. The fluctuations in abundance generated by normal rainfall variability are fairly minor (Fig. 8.2). The variability in the rainfall can be amplified to represent a more arid environment with wider fluctuations around the mean. In contrast, episodic disturbances occur at random intervals around some mean period. The survival rate of younger age classes is assumed to be somewhat more sensitive to disturbances, for example, 0.6 becomes 0.6 cubed, that is, 0.216. The reduction in survival is based on the effect that extreme cold spells toward the end of the dry season had on the study population of kudus (Owen-Smith 2000). Such disturbances have been set to occur every 10 years on average, but with a random number determining whether such an event occurs in any specific year (Fig. 8.3). The years when a disturbance occurs are indicated in the output on the screen. The impact of the disturbances is greater when they happen to coincide with years when the basic survival rate is lowered due to adverse rainfall conditions (Fig. 8.4).

This model is designed to allow you to explore the likelihood that a population of a specific size (restricted by the "saturation biomass") will persist over some time horizon, up to 200 years. You can investigate the effects of changing the population size, the stochastic influences incorporated, and the basic vital rate settings. Output in the form of the annual population structure, total size, and proportional growth rate is written to the computer screen. The population trend over time, as well as the density relationship, can be graphed, and the data used to produce these graphs

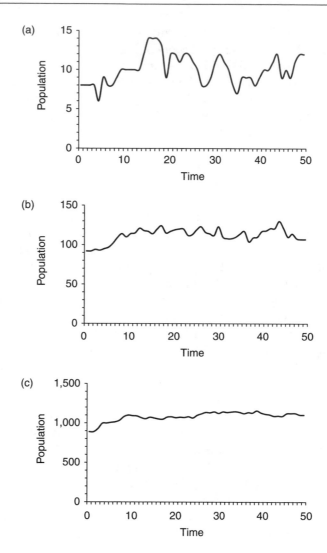

Figure 8.1 Output from model incorporating only demographic stochasticity, for three different population levels: (a) 10 adult kudu equivalents, (b) 100 adult kudu equivalents, and (c) 1,000 adult kudu equivalents.

saved as comma-delimited text, allowing retrieval into a spreadsheet for later viewing and analysis. In this model, extinction occurs when the population size reaches zero. There is no Allee effect reducing survival chances at low abundance.

The model is a reasonably realistic representation of the dynamics of the kudu population on which it was based. It can be adjusted to represent any

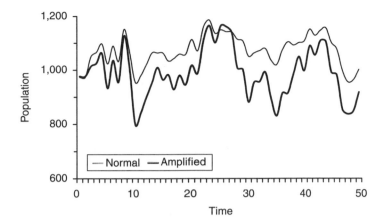

Figure 8.2 Modeled effects of rainfall variability on population abundance using real rainfall data with mean around 650 mm and coefficient of variation around 25%, compared with amplified rainfall variability achieved by squaring rainfall relative to the mean, giving a coefficient of variation of around 35%.

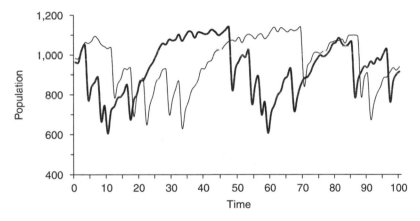

Figure 8.3 Modeled effects of episodic disturbances occurring at mean intervals of 10 years, and reducing survival rates when they occur, on population abundance with rainfall held constant, showing output from two runs with identical parameters.

basically similar species by altering the vital rates appropriately. A kudu can be changed into a moose by doubling the default fecundity from 0.5 (i.e., single calf produced annually) to 1.0 (typically twin offspring). Raising the fecundity automatically lowers the basic survival of the offspring. There is no further need to adjust the basic adult survival because under ideal conditions, a female moose should survive just as well as a female kudu. The basic survival rate of the juvenile and immature segments is automatically

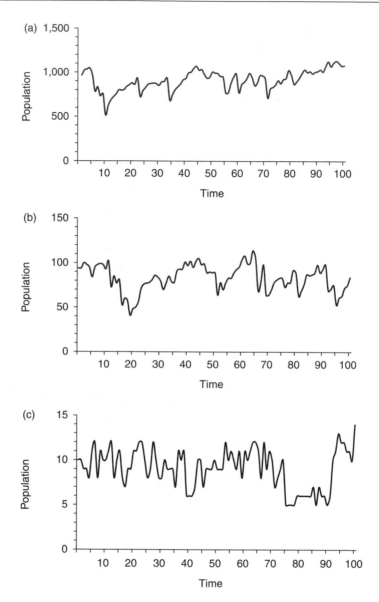

Figure 8.4 Modeled interaction between environmental variability and episodic disturbances, (a) suppressing demographic stochasticity by making the saturation biomass level 1,000 adults, and revealing demographic stochasticity by making the saturation biomass (b) 100 adults and (c) 10 adults.

lowered by a certain proportion relative to the adult survival rate. The density feedback reduces the survival of the adult segment at zero growth sufficiently to counterbalance the higher fecundity, as expected from life history theory.

Transform kudus into warthogs by setting fecundity to 2, that is, a litter of four piglets produced annually. Make the herbivore a rabbit by assigning a fecundity of 5 (i.e., 10 babies produced each year), and see what happens. Unrealistically, rabbits still take 2 years to reach reproductive maturity, unless you also change the "MinAdAge" setting near the top of the program code from 3 years to 1 year.

You can also use this model to investigate a very basic question: what is the chance that a founder population consisting of just two adult females (plus a male somewhere) will survive, even in an ideal constant environment? The initialization subroutine is set up so that these starting animals immediately produce offspring, hence this expanded total is the actual founding unit. Undertake repeated trials to assess the proportion of runs that result in extinction, that is, total population size falling to zero. You need not make the time span very long, because once the population increases much above the founding pair its chances of extinction become much slimmer. Ten years should be adequate. How many individuals should be introduced initially to ensure a greater than 95% probability of persistence?

There is an option to undertake repeated runs over the specified period, which can be activated by changing the "maxiterate" setting at the top of the program code to a value greater than one. However, iterations can be done more efficiently with a further elaboration of this program.

8.5.2 Assessing population viability

In any one run of the program SMALLPOP, the population may either persist or become extirpated before the end of the time period. In order to estimate the probability that the population will persist for this period, given the assumptions built into the model, you will need to conduct many repeated trials. To obtain an estimate with a precision of 1–2%, at least 1,000 runs will be required.

The program POPVIAB has been written to enable repeated trials to be conducted more efficiently. As in SMALLPOP, you can choose the forms of stochasticity included, and alter the default survival and fecundity rates, if you wish. In this version, there is also an option to make the annual variability in environmental conditions genuinely stochastic, through calling a random number ranging between 0.5 and 1.5. Alternatively, you can use the real rainfall records read in from the end of the program code, or amplify the variability based on these data. Each successive iteration commences with the population left at the end of the previous one, thereby effectively extending the period spanned until extinction takes place. After an extinction has occurred, the population is reset to the original starting size. In this model, a population consisting of just a single animal is interpreted as effective extinction. The annual population structure and total is printed on the screen during each run, and the occurrence of episodic disturbances is indicated. The output tells you when an extinction has occurred, both on the screen and through sound effects. It also displays a graph showing the change in population size during each run. To enable the screen output to be read, there is a "ViewDelay" setting at the start of the run. To make the program run faster, set the view delay to zero.

At the end of the set of iterations, there is an option to produce a new graph. This plots the persistence frequency as a function of time. This is the plot that you need to estimate the persistence probability over any portion of the specified time horizon. Even with 200 runs, the plot will appear somewhat wavy because of chance effects on when extinction occurred. You have the option to save the data generating this graph for later viewing and analysis.

The time duration selected for each trial does not matter much because probabilities can be multiplied, for example, the probability of a population persisting for a century is the probability that it will survive for the first 50 years times the probability that it will survive the next 50 years (i.e., the probability of persistence for 50 years squared). The main need is to get a reliable estimate of the probability. A hundred runs would still incorporate an uncertainty of several percentages around the estimate, which is quite great when the expected probability itself is only a few percent. A more precise estimate would require a thousand or more trials. The shorter the period spanned in each iteration, the greater the number of iterations required for a reliable estimate. The storage arrays are set to hold data from up to 200 iterations, for a time span of up to 200 years. Hence for 1,000 iterations, you will need to restart the program five times.

Using the default setting giving a saturation biomass equivalent to seven adults, the model estimates the probability that individual kudu herds of around this size will persist. As you will observe, the herd size can sometimes exceed this "carrying capacity," through chance effects, and of course the effective maximum biomass varies between years as a function of rainfall. Nevertheless, you will observe that each group will eventually fade to extinction, through stochastic processes. Given enough time, this could eventually lead to the extirpation of the whole population, if dispersal movements reestablishing groups in vacant home ranges became blocked. Having established the basic extinction risks for a kudu herd,

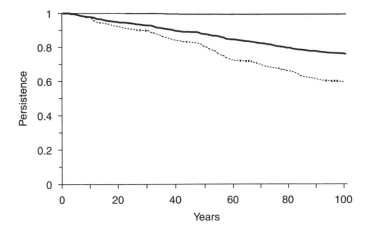

Figure 8.5 Comparative likelihood of persistence over increasing time horizons for (i) a large kudu herd of around 12 animals, taking into account episodic disturbances as well as demographic stochasticity for normal rainfall variability with a CV of around 25% (solid line), (ii) a typical kudu herd of around 7 animals for the same conditions (heavy solid line), and (iii) a kudu herd of around 7 animals under amplified rainfall variability with a CV of 35% (dotted line).

for the default population parameters, adjust the parameter settings to represent some other animal species. The program enables population viability to be assessed for any animal species that is roughly similar to a kudu in its environmental relationships.

Printing information to the screen slows the running of this program, even for a view delay of zero. The program PVAQUICK provides a more speedy tool for obtaining a final estimate of the persistence probability, by omitting the screen output, except for the starting situation. It allows up to 1,000 iterations, so as to obtain a sufficiently precise estimate of the persistence probability over time without having to repeat runs (Fig. 8.5).

8.6 Overview

The persistence probabilities derived from PVAs are only as good as the models used to generate them. As you have learned, all models entail simplifications and incorporate assumptions. The model output represents an optimistic estimate of the population viability because of various influences omitted. For instance, the model considers only the female segment. A sexually reproducing population is nonviable if no males survive, however many females remain, unless males can be introduced from elsewhere. Through this aspect alone the actual extinction probability might be double that estimated by considering females alone. The model ignores the effect of genetic inbreeding on offspring viability and fecundity when numbers remain small, and does not consider Allee effects reducing vital rates at low abundance.

For most rare species, we have little reliable information; therefore, we must formulate the model largely by extrapolation from similar species. Even for common species, little information is generally available about how survival probabilities change over time, or even what factors affect them. Future environmental conditions are unpredictable. Habitats may deteriorate in their capacity to support the species of concern, due to human impacts or climatic change. The main value of PVA models is to provide some objective assessment of the *relative* risks of population extirpation for different species and the comparative effectiveness of different interventions (McCarthy et al. 2001; Reed et al. 2002).

From this chapter, you should have learned the following:

1 Unless the population size is very large, demographic stochasticity causes the population trajectory to appear somewhat more erratic than the smooth sigmoid curve described by simple deterministic models.
2 The probability of population extinction rises rapidly as the population size becomes very small.
3 Demographic stochasticity alone rarely leads to extinction, unless the total population size supported is under about 10, or the expected population growth rate is close to zero or negative.
4 Some populations may be constituted by social units small enough to have a substantial probability of becoming extirpated.
5 Environmental variability can substantially increase the risk of population extirpation especially if the range of variation is large.
6 Episodic events with severe effects on survival rates can greatly influence on population persistence, especially when by chance two such events occur in close succession.
7 Species with high fecundity and hence high potential population growth rates do not necessarily have a higher viability than more slowly growing species, because high adult survival rates confer some resilience to disturbances.

Recommended supporting reading

The book by Burgman et al. (1993) provides a general overview of the pervasive effects of stochasticity on population dynamics, and population modeling generally. Morris and Doak (2002) give a definitive treatment of PVA, while Sjogren-Gulve and Eberhard (2000), and Beissinger and McCullough (2002) present chapters reviewing specific aspects. A condensed review is provided by Beissinger and Westphal (1998), while Fagan et al. (2001) assess factors affecting population vulnerability for a wide range of species. Lindenmayer et al. (1995) review various generic computer programs that have become available for modeling population viability. Critical appraisals of the shortcomings of PVA models are presented by White (2000) and Norris (2004). A population viability model for a plant population is described by Gotelli and Ellison (2006).

Programs on the accompanying CD

RANDWALK; SMALLPOP; POPVIAB; PVAQUICK

Exercises

1 Species with slow recruitment and hence recovery from low numbers are believed to be most vulnerable to extinction, but is this really true? Investigate this comparing the viability of similar-sized populations for herbivore species differing only in their fecundity, for example, a kudu (fecundity 0.5), a moose (fecundity $\simeq 1$), a warthog (fecundity $\simeq 2$), or a rabbit (fecundity 4–6).
2 McLoughlin and Owen-Smith (2003) suggested that elevated predation pressure lowering adult survival substantially reduced the viability of a roan antelope population, by restricting the recovery rate from episodic droughts. Evaluate this effect by lowering the default value for maximum adult survival in the PVA model.
3 Analyze how frequently animals would need to be translocated between isolated areas, in order to reestablish local small populations that had gone extinct by chance, for organisms with different social group sizes.
4 Assess the likelihood of persistence over a century of a kudu population remnant of under 20 animals, assuming (i) they form one cohesive unit, (ii) they consist of two separate herds occupying adjoining home ranges.
5 Compare the risk of extinction projected by the program POPVIAB with that obtained from other software packages available to you (e.g., VORTEX). Why do you think their projections differ? Which software would you choose, and why?

9

Metapopulation models

Spreading the risk

Chapter features

Topics

Dispersal and colonization; extirpation versus extinction; habitat patches; patch occupancy and incidence; correlated disturbances; sources and sinks; mainlands and islands

9.1 Introduction

The extirpation of a local population would not be a species extinction if other populations of the same species persisted. Moreover, the vacant habitat could later become recolonized by migrants from the populations that remained. The regional set of such local populations, interconnected by dispersal movements, is termed a *metapopulation* (Fig. 9.1). Just how much separation is needed between localized groupings of individual animals to regard them as components of a metapopulation is somewhat vague. Social units occupying distinct home ranges can become isolated by human occupation of much of the landscape. Clusters of these social units may be separated by unsuitable habitat, especially in mountainous terrain.

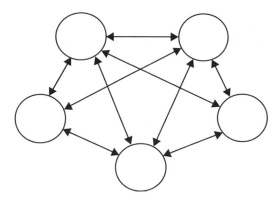

Figure 9.1 Schematic outline of a metapopulation, showing populations occupying isolated habitat patches connected by dispersal.

Lakes support distinct populations of fish and other aquatic organisms. Oceanic islands may be so widely separated from mainland populations that dispersal becomes a rare event.

Two important features need to be emphasized. First, all local populations have a finite chance of becoming extinct. Even if the local population is fairly large, extinctions can still occur through catastrophic events, or long-term shifts in habitat suitability. Second, dispersal movements are essential to reestablish populations that have faded or crashed to zero numbers, otherwise local extirpations lead progressively toward species extinctions. Hence the focus in metapopulation models is on the interaction between the chance of extinction of local populations and the rate at which the vacated habitat patches are subsequently recolonized. Rather than considering the local population abundance, our concern is merely whether a habitable patch supports a population or not.

Species that are widely distributed in many local populations have a reduced likelihood of becoming regionally extinct, through "spreading the risk." Recall that probabilities operate multiplicatively. Hence if the chance that any one population will be extirpated over some defined time period is 0.1, the probability that two populations will independently become extirpated over the same period is 0.01, and so on. However, local extinctions may not occur independently, depending on the regional prevalence of events with catastrophic impacts on population persistence, and factors governing dispersal movements following such occurrences.

With increasing fragmentation of landscapes by human expansion, many populations that were formerly continuously distributed have become broken into separated remnants, with the extent of the distribution also commonly contracted. Moreover, dispersal is often inhibited by hazards in traversing the human-transformed areas separating suitable habitat. Metapopulation models assess the risks of species extinctions as a consequence of such fragmentation, and identify how specific actions, such as providing dispersal corridors, can reduce such risks.

This chapter will outline some of the considerations involved in developing metapopulation models at specific scales, from the periodic demise of local social units every few years to extinctions and reestablishment of species in island archipelagos over millennia. You will receive guidance in the formulation of simple spreadsheet models to explore the dynamics generated.

9.2   Basic patch incidence model

The basic metapopulation concepts were formulated by Levins (1969) and subsequently revised by Hanski (1994, 1997) and others. Let P be the proportion of sites occupied by populations, E the expected rate of local extinction of these populations, and C the expected rate of colonization of vacant sites by migrants from occupied patches (with the label "expected" implying that these processes are *stochastic*). The time dimensions for these rates will generally span several years or decades depending on the organism being considered. The change over time in the proportion of patches occupied depends on the balance between colonization and extinction, that is,

$$\Delta P / \Delta t = CP(1 - P) - EP. \tag{9.1}$$

Note that the effective colonization rate depends on the proportion of patches containing populations. Provided the colonization rate exceeds the extinction rate, the proportion of occupied patches converges toward an equilibrium, just as the logistic equation projects an equilibrium population size based on the balance between births and deaths (the two equations are functionally identical). This equilibrium proportion P^* is given by

$$P^* = 1 - E/C. \tag{9.2}$$

This means that some sites containing suitable habitat will not support a population at a particular time, as a result of the chance processes of local extinction considered in Chapter 8. Patch occupancy will become zero if the extinction rate exceeds the colonization rate.

However, if there is a substantial "propagule rain" from some large "mainland" population, colonization may be independent of the occupancy of the remaining "island" patches. In this case, the equilibrium proportion is given by

$$P^* = C/(C + E). \tag{9.3}$$

Hence, at equilibrium some patches may be occupied even when the extinction rate exceeds the colonization rate, because of the constant propagule rain into vacant habitat. This starting model can be refined further to consider differences in patch or island size, affecting the extinction probability, and in distance from the mainland source, affecting the colonization probability (Hanski 1994, 1997).

Metapopulation dynamics can be modeled easily in a spreadsheet, since we are dealing with a fixed number of patches. Follow the guidelines in Appendix 9.1. The output represents whether each patch is occupied or vacant and the changes in this state over some time step. Start with different initial proportions of occupied patches, and observe how over time the proportion converges toward the value for P^* predicted by eqn (9.2) (Fig. 9.2). Due to the stochastic formulation, the proportion observed in any time-limited trial may deviate somewhat from that ultimately expected. Alter the values of E and C, and observe how this changes the eventual proportion of patches occupied. When these two rates are equal, regional extinction of the metapopulation is inevitable, but may take a long time if the extinction rate is low. You can establish this by chaining successive trials, using the end state from each as the starting state of the next, or by increasing the rates to represent a lengthened time step. Reduce the number of habitat patches, and note how this raises the chance of regional extinction for the same parameter values.

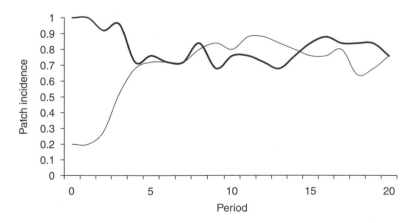

Figure 9.2 Convergence of patch incidence toward the proportion predicted by the relationship between colonization and extinction rates, showing the output from two trials started with different initial proportions.

This basic model assumes all patches to be identical, so that the populations they support have identical chances of extinction. Moreover, extinctions occur independently because the outcome is determined by a separate random number for each patch. The probability of being recolonized is also the same for each habitat patch, suggesting that these patches are equidistant. A further assumption is that the basic probabilities of extinction and colonization do not change over time. These conditions can be altered to represent some more realistic situation.

9.3   Correlated migration and extinction

In the model you developed above, recolonization occurred independently in each habitat patch, that is, a different random number was used to decide whether or not vacant patches became reoccupied in each time step. It might be more reasonable to assume that whether migrants set out from occupied patches depends on the environmental conditions in the region, common to all patches. For example, dispersal tends to occur just before or after the population level that can be supported is exceeded (Lidicker 1962, 1975). Furthermore, migrants may not make it through the intervening matrix at times when conditions are adverse, for example, lack of water or food.

To represent this situation, a single random number applied to all patches determines whether recolonization occurs in each time step. Specifically, if the colonization rate C is set at 0.5, all vacant patches get reoccupied during iterations when the random number is less than this value, while no migration occurs in other time steps. This change raises the likelihood of metapopulation extinction due to the progressive reduction in the number of occupied patches during periods when no migration occurs, reducing the number of migrants eventually sent out when conditions become favorable. Modify your model accordingly, following the guidelines in Appendix 9.2, and observe the difference made for specific values of C and E.

The likelihood of extirpation of local populations could also depend on climatic or other conditions prevailing across the region, in particular, the occurrence of catastrophic events such as droughts or floods. However, modeling such "correlated disturbances" cannot be done quite as simply as for migration, otherwise all of the local populations blink out in the same step when the random number happens to fall below the extinction probability. Instead, we need to conceptualize the environmental disturbance as increasing the likelihood of local extirpation above some background risk due to local processes within each population. This allows for the possibility that certain lucky populations could survive these events when they occur, and thus be a source of recolonization. Representing such a situation is

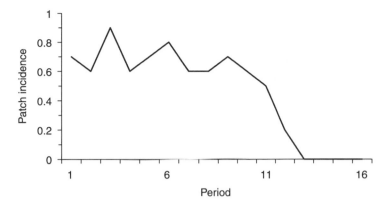

Figure 9.3 Model output showing how a correlated disturbance can result in the sudden collapse of a metapopulation.

described in Appendix 9.2. Correlated disturbances increase the chance that the entire metapopulation will become extinct, for the same probability settings for patch extinction and colonization. Moreover, regional extinction can come about quite suddenly, when a disturbance happens to leave so few surviving populations that recolonization chances are greatly reduced (Fig. 9.3).

9.4   Variable patch size and spacing

In real-world situations, habitat patches vary in their size and spacing. Large patches support large populations, which are unlikely to go extinct through chance demographic processes. Remote patches are less likely to be recolonized when they become vacant. The context could take the form of a large core population, surrounded by several small peripheral populations (Fig. 9.4). The latter may suffer regular extirpations, but be restored from the core population, with the most remote patches being vacant for longest. However, there is a possibility that even the large core population could occasionally be wiped out by some extreme catastrophe, such as a disease outbreak or a massive flood. By chance one of the peripheral populations could survive the event, and eventually recolonize the core habitat patch. Hence there is a lessened chance of total extinction for species that are widely distributed in discrete populations.

Appendix 9.3 describes how to modify the spreadsheet model to represent this situation. Undertake repeated trials using some basic parameter values to see how chance alone can produce widely divergent outcomes. Note how

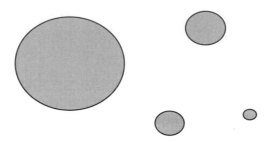

Figure 9.4   Metapopulation structure in the form of a large core population and several peripheral populations isolated by different distances.

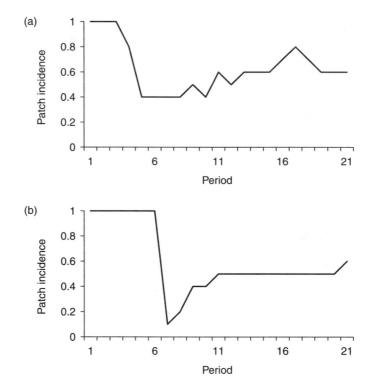

Figure 9.5   Model output showing recovery of a metapopulation after a major disturbance (a) through persistence of the core population and (b) through persistence of one of the peripheral populations.

correlated disturbances affecting the entire habitat patch mosaic sometimes cause the collapse of the core population along with all peripheral populations, but very occasionally one of the peripheral patches happens to survive and slowly reestablishes a population in the core patch (Fig. 9.5).

Recolonization is slow in such circumstances because the number of migrants departing from the peripheral patch is small. It is also dependent on luck, through not having the peripheral population wiped out by some further disturbance, or chance effects, before the core population has been reestablished.

9.5   Source and sink populations

Localized populations may also be differentiated in terms of their intrinsic growth potential. Patches of favorable habitat support dense populations, restricting opportunities for new recruits to settle. Hence newly maturing animals or birds disperse in search of less crowded conditions where they can perhaps establish a territory and breed. The only vacancies may lie in marginal habitat where expected mortality losses exceed the likely reproductive success. Populations in these "sink" patches are maintained by emigrants from the core, or "source" habitat.

Set up such a model following the guidelines in Appendix 9.4, with the migration rate set at one, meaning that dispersal always occurs from the source habitat, and even from sink populations if these exist (animals from these patches seeking better conditions). Make the extinction rate equal to or greater than the effective colonization rate in the sink habitats, but with the source population invulnerable to extinctions. You should find that populations can be maintained in sink habitat for prolonged periods.

From a conservation standpoint, concern lies in what happens if the source population were to be eliminated. This could occur through its habitat being preempted for human use, or through the population being wiped out by some natural catastrophe. The source population could be reestablished through emigration from sink populations, provided that these persist. However, if the conditions in the source habitat no longer support a population there, the metapopulation remnant in the sink patches rapidly fades out. The message: many populations persist only through their connections to migrants from more favorable but crowded habitats.

9.6   Mainland–island habitats

For populations isolated on genuine islands, in addition to the size of the island the distance from the mainland affects recolonization prospects. Even some bird species avoid crossing water barriers. Hence islands inevitably support fewer species when compared with the nearby mainland. Similar principles apply to isolated terrestrial habitats within a matrix of inhospitable terrain, such as protected areas surrounded by human habitation.

This is the situation modeled in island biogeography (MacArthur and Wilson 1963, 1967), with a focus on how species diversity is affected by island size and isolation. The time frame is now greatly lengthened. In many situations, the equilibrium species number will not yet be reached because insufficient time has passed since the island became isolated from the mainland. The relevant time periods for many organisms may span centuries or even millennia. Refer to Gotelli (2001) or Case (2000) for suggestions about how to model species–area relationships as a function of immigration and extinction rates.

9.7   Examples of vertebrate metapopulations

Most metapopulation models have been applied to short-lived invertebrates like butterflies (Hanski et al. 1994; Thomas and Hanski 1997). Nevertheless, surveys of distribution patterns can provide some evidence of metapopulation processes affecting larger vertebrates.

For example, the midlands region of the KwaZulu-Natal Province of South Africa contains forest patches varying in size within an open grassland matrix. Certain mammal species are restricted to these forest patches, including a monkey (*Cercopithecus mitis*), a small antelope (*Philantomba monticola*), and the tree hyrax (*Dendrohyrax arboreus*) (Lawes, Piper, and Mealin 2000). Many of the forest patches are smaller than the typical home range size of these species: 45 ha for the monkey, 0.7 ha for the antelope, and probably somewhat smaller for the hyrax. Nevertheless, these mammals were frequently absent even from patches above this size: from 30% of forests greater than 50 ha for the monkey, and 55% of those greater than 1 ha for both the antelope and the hyrax. Some of the forest patches may not have been suitable habitat. Moreover, occupancy decreased with increasing distance from the nearest occupied forest for the antelope and hyrax, which were not found in any forest isolated by more that 1.5 km. The monkey seemed not to cross even smaller gaps.

Spatially explicit modeling of metapopulation dynamics has been conducted in the context of conserving the spotted owl (*Strix occidentalis*), restricted to old-growth forest patches in the western coastal region of the United States and Mexico (Gutierrez and Harrison 1996; Lamberson et al. 1994; Noon and McKelvey 1996). Harrison and Taylor (1997) review the empirical evidence for metapopulation dynamics more widely.

9.8   Overview

In this chapter you should have learned the following:

1 Local population extinctions lead inevitably to regional species extinctions if dispersal movements are blocked or strongly inhibited.
2 Some patches of suitable habitat may not contain populations because of chance local extinctions.
3 Regionally correlated disturbances have a large impact on metapopulation persistence.
4 Some populations may persist in intrinsically unsuitable habitat through continuing immigration from areas of favorable habitat.
5 Eliminating a source population through habitat transformation can lead to the collapse of the regional metapopulation.
6 Isolated islands support a reduced species assemblage resulting from the balance between local extinctions and recolonization from a mainland area.

Recommended supporting reading

Succinct reviews of metapopulation theory are given by Gotelli (2001) and Hanski (1994). For a fully comprehensive treatment of metapopulation theory, see Hanski (1999). Examples of spatially explicit simulations are provided by Lamberson et al. (1994) for spotted owls, and by Thomas and Hanski (1997) for butterflies. An application of the incidence function model to shrews is described by Hanski (1992).

Exercises

1 Investigate how reductions in the number of populations constituting the metapopulation affect the likelihood that the regional metapopulation will persist for different combinations of extinction and colonization rates.
2 Establish the extent to which regionally correlated disturbances affecting the likelihood of local extinctions reduce the overall persistence probability of the regional metapopulation.
3 Transform the source–sink metapopulation model into a mainland–island model, and establish how the number of populations each representing a different species declines with increasing isolation from the mainland, and consequently lowered colonization rate.

10

Modeling infectious diseases

Outbreak dynamics

Chapter features

Topics

SIR model; transmission rate; basic reproductive number; threshold population size; mass action; frequency dependence; control measures

10.1 Introduction

The threat that disease outbreaks can pose for conservation has recently become widely recognized (McCallum and Dobson 1995; Daszak, Cunningham, and Hyatt 2000). Canine distemper caused the demise of the last wild colony of black-footed ferrets (*Mustela nigripes*; Thorne and Williams 1988), while similar morbilliviruses have infected African wild dogs (*Lycaon pictus*; Gascoyne et al. 1993), Ethiopian wolves (Sillero-Zubiri, King, and MacDonald 1996), lions (*Panthera leo*; Roelke-Parker et al. 1996), and seals (Heide-Jorgensen and Harkonen 1992), with severe mortality losses. The morbillivirus causing rinderpest (the bovine counterpart of measles) was accidentally introduced into Africa in the late nineteenth century, and within a decade had swept across the continent killing over 90% of buffalo and other ungulate species (Plowright 1982). Rinderpest persisted in the Serengeti wildebeest population until the 1960s as "yearling disease," causing almost 50% mortality among young animals. It disappeared from wildlife populations following the vaccination of cattle. Anthrax (a bacterial disease) has exhibited recurrent outbreaks among wild ungulates in South Africa's Kruger National Park (Bengis, Grant, and de Vos 2003) and causes regular mortality even among elephants in Etosha National Park in Namibia (Lindeque and Turnbull 1994). Bovine tuberculosis has spread from buffalo to affect the lion population in the Kruger National Park, and hence threatens tourism in the region (Bengis et al. 2003).

Protozoa, bacteria, and viruses living entirely within other organisms, which sometimes become pathogenic (disease causing), are generally labeled *microparasites*. *Macroparasites* in the form of tapeworms and roundworms may also have adverse effects on host populations, most notably contributing to the cycling of red grouse (*Lagopus lagopus*) in Scotland (Hudson, Dobson, and Newborn 1998). The focus in this chapter will be mostly on microparasites causing infections. Numerous questions are raised. Why do some microbial parasites persist in populations whereas others generate only transient outbreaks of infection? Why do outbreaks eventually fade out before the host population has become completely infected? How great an effect do parasites have on the abundance of their hosts? How can diseases threatening conservation objectives best be controlled?

10.2 Basic infection model

Models of disease dynamics focus on the proportion of the host population infected, rather than changes in the parasite population. They are similar in form to models of metapopulation dynamics, except that more than two patch states are distinguished (Fig. 10.1). *Susceptible* hosts become *infected*, and after a period either *recover* from the infectious state, perhaps with long-lasting immunity, or die. Accordingly, this is labeled the "SIR" model. In more elaborate models, an incubation period during which the host is no longer susceptible to further infection, but cannot yet pass on the pathogen, is also distinguished.

Figure 10.1 Diagrammatic outline of the SIR model.

A disease spreads as a result of contacts between infected and susceptible hosts, through "mass action." The spread is counteracted by the rate at which host individuals lose their susceptibility, by acquiring immunity or dying. Hence the change in the number of infected hosts over time can be described by the following dynamic equation:

$$dI/dt = \beta SI - \nu I, \tag{10.1}$$

where I represents the number (or density) of susceptible hosts, S the number (or density) of susceptible hosts, β the transmission rate, and ν the rate of recovery of infected hosts. The transmission rate depends on the probability that a contact will lead to an infection, together with the likelihood that such close contacts will occur in relation to the effective population size. The product βI has been labeled the "force of infection."

Whether the pathogen can initially spread in the host population is a crucial issue. The crux is whether each initial infection produces more than one secondary infection, leading to a measure termed the basic reproductive number or ratio of the disease R_0. By definition, R_0 represents the number of secondary infections created by each primary infection in a completely susceptible host population. Note that there is no time dimension. The crucial issue is whether or not the disease spreads, not how rapidly it spreads. If each new infection leaves less than one subsequent infection, the disease fades out. The basic reproductive number is the product of the effective transmission rate, the duration of the infection, and the number of susceptible hosts, which initially is the total population size N. The duration of the infection is the inverse of the rate of recovery from the infection. Hence

$$R_0 = \beta N/\nu. \tag{10.2}$$

Note how the temporal dimension of β and ν cancels out. Hence diseases tend to spread if they have either a high infection rate or slow recovery (long duration of infection), or some appropriate combination. There is also a threshold population size or density N_T needed for the disease to spread. This is obtained by rearranging eqn (10.2) and setting R_0 equal to 1, yielding $N_T = \nu/\beta$.

The spread of infections causes the number of susceptible hosts to decline. Those infected either recover and lose their susceptibility through becoming immune, or die, and in either case no longer transmit the disease. The decrease in the proportion of the population that is susceptible eventually causes the effective reproductive number, given by the product $\beta SI/\nu$, to fall below one, with the result that infections fade out. A new outbreak occurs when infectious individuals enter or remain in a population that has become largely susceptible in the absence of infections. This is the typical outbreak cycle shown by transient infections.

For some diseases, transmission depend on the *proportion* of susceptible individuals within the population, rather than on their absolute number or density. In other words, spread is *frequency dependent*. This is the case for sexually transmitted diseases. Since contacts are actively sought, rather than happening randomly, contagion can continue to occur even in small populations. Proportional dependency could also arise in a very dense population, where the rate of contact saturates. The prevalence of brucellosis, causing abortion, among bison in Yellowstone Park seemed to be most consistent with this pattern, perhaps because the animals occur in large herds (Dobson and Meagher 1996).

The formula for the reproductive number then becomes

$$R_0 = (\beta/\nu)(S/N) \tag{10.3}$$

and in a completely susceptible population $S/N = 1$. Hence initial spread of a sexually transmitted disease depends simply on whether the transmission rate β is greater that the recovery rate ν, or exceeds the death rate if there is no recovery. As more individuals acquire the disease, the proportion S/N declines, to a stage in which further spread is halted. What happens next depends on whether the pathogen can persist – in a dormant form, another host, or somewhere in the environment. Spatial variation in the proportion infected and temporal variation in the rate of transmissions can contribute toward maintaining infections, unless host mortality removes the infective agent sufficiently rapidly.

For this chapter, two simple models will be developed, one applicable to a highly infectious virus producing a transient infection but no mortality, and the other for a sexually transmitted virus with slow transmission causing eventual death of the host. How to model the efficacy of potential measures to control the spread of a wildlife disease will also be outlined.

10.3   Cyclic outbreak dynamics: measles

The dynamics of measles infections in human populations have been particularly well studied (Fine and Clarkson 1982; Bjornstad, Finkenstadt, and Grenfell 2002). In the past, measles showed a regular 2-year period between outbreaks among school children in most large cities around the world. This pattern faded out after widespread vaccination against the disease was introduced in the 1970s (Fig. 10.2). Hence four interesting

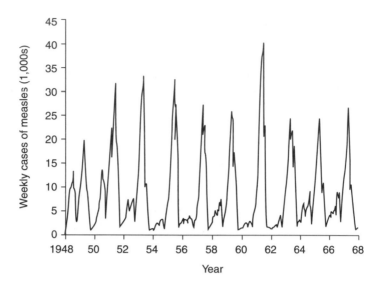

Figure 10.2 Cyclic outbreak dynamics of human measles shown in England and Wales and later suppression by vaccination (from Begon et al. 2005).

questions arise:

- What initiated the outbreaks?
- What caused the outbreaks to fade out?
- Why was there a regular 2-year period between outbreaks?
- How did vaccination cause the outbreaks to cease?

The model you will develop applies to the period prior to vaccinations (Appendix 10.1). New infections spread rapidly in an initially completely susceptible population. Following a latent period of about 10 days, the infectious period during which the pathogen can be passed on spans 7 days. By the time children leave school, almost all will have had measles, and recovered with life-long immunity. Very few children die from the disease, at least in first-world countries with good medical care.

The effective value for β in the model is contingent upon the measure of population size. The disease spreads initially only if the product $\beta N/v$, representing R_0, exceeds one. Infections start declining once the effective reproductive number, that is, the product $\beta SI/v$, falls below one. By this stage a high proportion of contacts will involve individuals who are no longer susceptible to infection, having recovered from the disease. The number infected tends inevitably toward zero because each infection now contributes less than one new infection. Nevertheless, at the time when

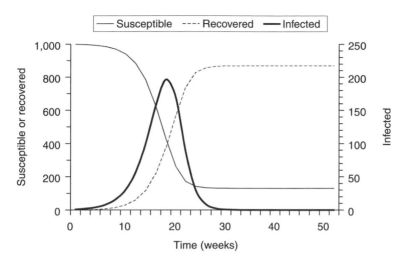

Figure 10.3 Modeled course of a transient disease outbreak showing changes in the number of susceptible, infected, and recovered individuals within the population.

infections fade out, many individuals still remain susceptible (Fig. 10.3). The higher the rate of transmission, the smaller the proportion remaining susceptible by this stage.

The cycle of infections can be sustained if new individuals enter the susceptible category, through births and subsequent commencement of schooling. This restores the number of susceptibles toward the level at which a new or persisting infection can spread, initiating a further outbreak. The outcome is a pattern of repeated peaks in infections, with the time between peaks depending jointly on the transmission rate and the "birth" rate into the susceptible population (Fig. 10.4a). However, the amplitude of the oscillations tends to decline over time. In reality, new scholars tend to arrive in a pulse at the start of each school year. This influx of susceptible children may cross the critical threshold, initiating an abrupt peak in infections. After these children have recovered, the number in the susceptible category may still be below the threshold size for infections to spread the following year. With a suitable combination of parameter values, the observed 2-year period between measles outbreaks can be closely replicated (Fig. 10.4b).

Vaccination reduced the proportion of children entering schools who are susceptible to acquiring the disease. The proportion that needs to be immunized in order to halt spread by reducing the effective reproductive number below one is given by $1 - 1/R_0$. The actual R_0 for measles is around 13, so that outbreaks became suppressed once 92% of children

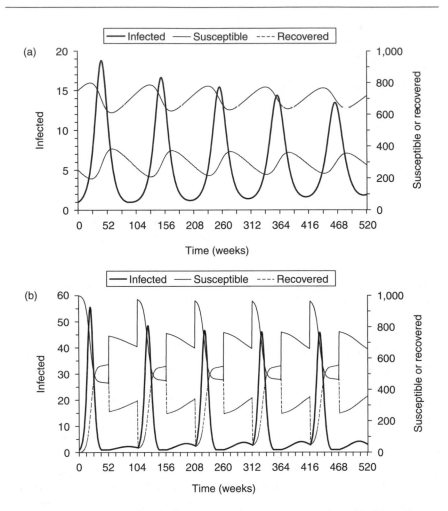

Figure 10.4 Cyclic outbreak dynamics with a 2-year period modeled by allowing either (a) continuous births into the population or (b) a pulse of births at the start of each year.

had received vaccinations. Nevertheless, measles has not been completely eradicated because infections still persist in places where vaccination is less complete.

Measles has another characteristic feature: it persists only in cities with populations exceeding around 300,000 people. The lower frequency of contacts in smaller towns means that infections are less likely to spread, while the chance of some infections persisting to initiate a new outbreak is also reduced. In contrast, the virus causing smallpox was eradicated by widespread vaccination. This is largely because R_0 for smallpox is only around 3–5, so that vaccinations needed to cover only 75–80% of the population in this instance.

More realistic models of measles dynamics need to take account of the spatial structuring of the population, as well as the pattern of spread among different age classes in the school-going population. Many other childhood diseases also showed repeated outbreaks in the past, but without the regular periodicity of measles. This might be because they are less infectious, or have a longer duration of infectiousness. Explore this using the model that you developed for measles.

As noted above, rinderpest is the bovine equivalent of measles, and probably the original source of the virus that crossed to humans when cattle herding became widespread. The transient wave of infections spreading across Africa that it showed is typical of viral infections conferring life-long immunity, despite the high mortality in this case. The persistence of the disease among wildebeest in the Serengeti resulted from repeated infections from cattle. Only a few isolated outbreaks have occurred since immunization of cattle became widespread. Phocine distemper affecting seals in the North Sea, likewise caused by a morbillivirus, showed a similar massive outbreak associated with high mortality, but has not recurred (Fig. 10.5, Grenfell, Lonergam, and Harwood 1992). Model projections indicate that the threshold population size for distemper to be maintained is vastly greater than the size of any existing seal colony, so that the virus must have crossed over from some other host species (Swinton et al. 1998).

Low virulence or weak immunity, or a long incubation stage, can lead to the pathogen persisting as an endemic infection. A biological reservoir in another species resistant to the effects of the pathogen can also be the source of new infections. Certain bacterial diseases, including anthrax, can survive for extended periods as dormant spores in soil.

10.4   Slowly spreading sexually transmitted disease: HIV–AIDS

Another human disease that has attracted much attention recently is the Human Immunodeficiency Virus (HIV) causing the Acquired ImmunoDeficiency Syndrome (AIDS; Williams and Campell 1996). Its characteristics are very different from those of measles. It is transmitted almost solely through close sexual contacts. Without treatment, infections are almost invariably fatal, with the median time between infections and death estimated to be around 7 years in Uganda, which has provided some of the best data. No immunity is acquired (with possible rare exceptions). The transmission rate is very slow, with each HIV carrier infecting another

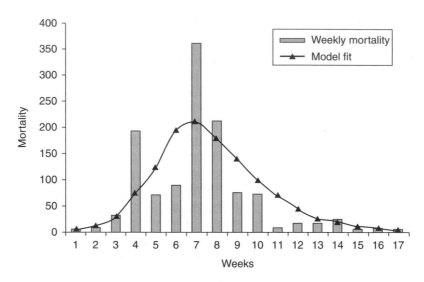

Figure 10.5   Dynamics of an outbreak of phocine distemper in a seal colony in the North Sea, as shown by weekly number of dead seals, compared with outbreak pattern projected by the model (after Hudson et al. 2002, following Grenfell et al. 1992).

person only every 1.25 years. Nevertheless, during the extended infectious period each initial carrier will infect $(1/1.25) \times 7$ others, so that R_0 is 5.6. Accordingly, once introduced, the virus spreads slowly, but inexorably, until effective interventions take place.

When AIDS came to the attention of mining companies in South Africa in 1987, a survey revealed that the HIV prevalence among mine workers was only 0.1%, concentrated largely in migrant workers from Malawi. Hence Malawians were repatriated and nothing further was done. By 1998 the prevalence had reached 20–30% among mine workers as well as among rural women attending antenatal clinics (Williams et al. 2000b). The failure to act was partly because human intuition tends to project an arithmetic sequence, that is, 1%, 2%, 3%, 4%, ... rather than a geometric series, that is, 1%, 2%, 4%, 8%, ... An appropriate model could have alerted mine owners to the impending catastrophe at an early stage, when intervention would have been most effective. Of course, at that time few data were available to construct a realistic model. Furthermore, the long delay between infection and death led to procrastination by government officials skeptical of the link between the virus and the disease, especially as HIV sufferers die largely from secondary infections.

You have the basic facts needed to construct an elementary model, following the guidelines in Appendix 10.2. The model represents the spread of HIV infections among sexually active adults aged between 16 and 49 years, taking no precautions. Starting with an initial prevalence of 0.1% in 1987, it projects a proportion infected rising to between 17% and 35% by 1999, depending on assumptions made about the lag between infection and death, and about the intrinsic birth and death rates in the population (Fig. 10.6). The actual prevalence of HIV infections among women attending antenatal clinics in rural areas of South Africa had exceeded 20% by 1998, while among male mine workers it had reached almost 30% (Williams et al. 2000a,b). Even higher prevalence levels of 60–80% were observed at that time in the population segments most exposed to HIV infections, that is, sex workers in mining areas and male truck drivers. AIDS-related mortality still remained low, simply due to low incidence of infections 7 years earlier.

Based on the rising prevalence, deaths as a consequence of AIDS were projected to rise to 3% of the susceptible South African population, or almost 0.6 million annually from an adult population of 20 million, by 2006 (Fig. 10.7). The spreadsheet models indicate a somewhat higher mortality loss of around 11% by this time, from the continuing steep rise in the projected prevalence of HIV after 1998. The actual prevalence of HIV among women attending antenatal clinics had leveled off at around 27% by 2004, while the overall prevalence among adults aged 16–49 years in South Africa had become 16%. The annual rate of infection at this time was estimated to be over 6% per year among young adults. The number of individuals infected with HIV projected by the model formulations peaks around 2004 when the prevalence is close to the theoretical maximum of 82% expected for an R_0 of 5.6, and is most closely approximated by the version assuming that death occurs exactly 7 years after infection. However, the proportion infected continues to rise thereafter in all models, because those surviving constitute a greater proportion of the diminishing population and keep passing on infections to the diminishing fraction remaining susceptible. The eventual proportions depend quite strongly on the intrinsic birth and death rates assumed as well as the pattern of survival after infection.

The model can be used to explore the likely consequences of various interventions, for example, the use of condoms reducing the rate of transmission of HIV by some fraction of the sexually promiscuous population. Other possible interventions include the treatment of sexually transmitted diseases, such as syphilis and gonorrhea, which greatly increase the risk of HIV transmission (Williams et al. 2000c; Gilgen et al. 2001). On the other hand, prolonging the life of HIV carriers could potentially increase the spread of infections because each individual has more opportunity to pass on the virus, unless sexual behavior changes. Fortunately, antiretroviral drugs greatly reduce infectiousness. More sophisticated models indicate that a

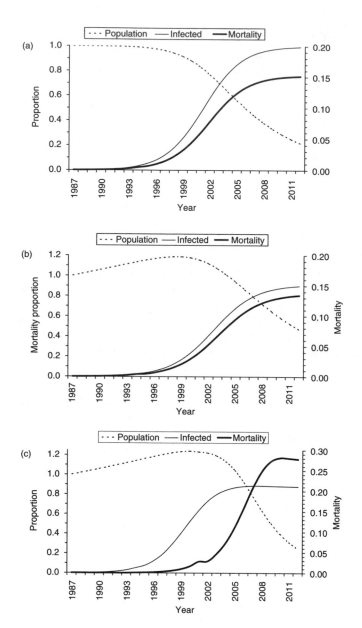

Figure 10.6 Modeled dynamics of HIV infections in the high-risk adult population showing changes in the projected total population size, proportion infected, and annual mortality rate assuming (a) widely variable survival, once infected, around a mean of 7 years, with zero births and no deaths from other causes, (b) widely variable survival around a mean of 7 years, with an annual birth rate into the susceptible segment of 4% per year, and annual death rate from other causes of 2% per year, and (c) deaths occurring exactly 7 years after acquiring an HIV infection, with birth and death rates as in (b).

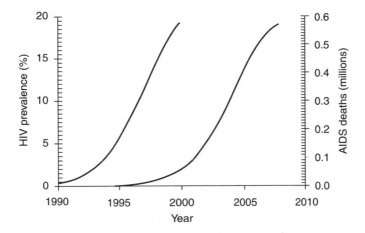

Figure 10.7   Projected rise in mortality from HIV–AIDS in the South African adult population based on the observed rate of increase in the prevalence of HIV infections assuming 7 years between infection and death (from Williams et al. 2000b).

few promiscuous individuals have a disproportionate effect on the spread of infections, so that interventions targeted at such "super-spreaders" could make the biggest difference (Lloyd-Smith et al. 2005a).

10.5   Controlling the spread of wildlife diseases

Diseases in wildlife populations that can be passed on to humans, their domestic animals, or endangered wild species have become a source of much concern (McCallum and Dobson 1995). These diseases will usually be endemic within the host population. Hence natural immunity is inadequate to eliminate the pathogen, while the pathogen causes relatively little mortality. However, severe mortality can result, or at least debilitation affecting survival prospects, when a secondary less-resistant species is infected. Notable examples include

1 rabies, transmitted from foxes in Europe and raccoons in North America to domestic dogs and sometimes humans (Anderson and May 1982; Jenkins, Perry, and Winkler 1988);
2 bovine tuberculosis, transmitted from badgers in Europe and possums in New Zealand to domestic cattle, and potentially to humans in countries where milk is unpasteurized (Clifton-Hadley et al. 1995; Roberts 1996);
3 variants of avian influenza ("bird flu") passed on via domestic fowl to humans.

Control rests on the concept of a critical population size or density for the spread of the disease in the source host. Some modification of the formula for R_0 is needed. Since there is no immune state, we consider the transition rate from susceptible hosts to a latent state harboring the pathogen, which at a later stage becomes infectious. We also need to allow for the birth rate into the susceptible category, as well as the death rate both from infections and from other causes. Hence the equation for the basic reproductive number, from Anderson and May (1981) assuming density-dependent mortality, becomes

$$R_0 = \beta K \gamma / [(\gamma + b)(\alpha + b)], \qquad (10.4)$$

where β is the transmission rate for the disease, γ, the rate at which latent carriers become infectious, b, the maximum birth rate, α, the additional mortality rate due to the disease, and K, the equilibrium population size maintained in the absence of infections. Hence the critical population size for the effective reproductive number to become one is

$$N_T = [(\gamma + b)(\alpha + b)]/\beta \gamma. \qquad (10.5)$$

Managerial interventions to halt or reverse the spread of a disease could entail either (i) reducing the abundance of the host species to below N_T, through some form of culling, or (ii) reducing the susceptible proportion below the threshold for spread, via vaccination. Equation (10.4) assumes transfer of the pathogen through random contacts between infected and susceptible hosts, that is, mass action dependent on population density. If transmission depends on the *proportions* of infected and susceptible hosts (i.e., frequency dependence), culling would obviously be ineffective and only vaccination (or sterilization) would be applicable. Vaccination would have to be done repeatedly to counteract the rate of dilution of the immune animals through births and deaths. Culling would also need to be repeated to counteract births.

The magnitude of the control action needed can be estimated approximately by setting N_T/K equal to $1/R_0$, leading to the following equations (from Barlow 1996):
Culling rate needed:

$$r[1 - (1/R_0)] \qquad (10.6)$$

with r representing the intrinsic population growth rate;
Vaccination rate needed:

$$b(R_0 - 1). \qquad (10.7)$$

Note that all of these rates are expressed as a fraction of the host population per unit time.

To establish the relative efficacy of alternative control measures, we need to model the basic population dynamics of the host species, differentiating also changes in the infectious and noninfectious segments within this population. Control actions based on the above equations can then be applied. Following Barlow (1996), and still assuming density-dependent mortality, we need three linked equations:

$$dX/dt = X(b - d - r(X/K)^{\theta}) - aY, \tag{10.8}$$

$$dY/dt = \gamma Z - Y[\alpha + d + r(X/K)^{\theta}], \tag{10.9}$$

$$dZ/dt = bY(X - Z - Y) - Z[g + d + r(X/K)^{\theta}], \tag{10.10}$$

where X is the total host density, Y, the density of infectious animals, Z, the density of hosts harboring the pathogen in a latent form, d, the minimum death rate, r, the intrinsic rate of increase (i.e., $b - d$), K, the equilibrium population density in the absence of the disease, θ, the shape parameter determining the form of density dependence, and other symbols are as for eqn (10.5).

Set up these equations in a spreadsheet and assess the consequences of different values of β and γ for the magnitude of the intervention required, whether through culling or vaccination. Evaluate Barlow's (1996) finding that culling is more effective than vaccination when R_0 for the disease is high (>3), while either culling or vaccination could be effective for lower R_0 values, depending on features of the host dynamics, such as the intrinsic growth rate.

In practice, the rate of spread of the disease may not be linearly dependent on the contact rate between infected and susceptible animals, as assumed in the mass-action model. If contact rates tend to saturate toward high density levels, culling becomes less effective. The frequency-dependent model of transmission developed for sexually transmitted diseases may be more appropriate under such conditions (Dobson and Meagher 1996). The number of animals within some neighborhood region determines the actual frequency of close contacts leading to transmission. Accordingly, the spatial dispersion of the host population is important as well as the overall population density. Furthermore, threshold densities for disease transmission are rarely abrupt because of spatial and temporal variability in contact rates within the population, and in practice difficult to estimate anyway. Despite these caveats, management interventions to reduce the abundance of susceptible hosts could still be effective in disease control (Lloyd-Smith et al. 2005a).

10.6   Overview

This chapter has provided some of the conceptual foundations underlying the broad and expanding field of ecological epidemiology. In this chapter, you should have learned the following:

1 The initial spread of a pathogen within a host population depends on whether the basic reproductive number is greater than or less than unity.
2 Spread is eventually halted when the proportion of susceptible individuals is reduced sufficiently so that the effective reproductive number drops below unity.
3 Many highly infectious diseases show outbreak dynamics with the pathogen disappearing from the host population unless there are sources of re-infection.
4 There may be a critical size or density of the host population for the initial spread of the disease.
5 For sexually transmitted diseases and certain other situations, spread depends more on the proportion of the population susceptible, allowing the pathogen to persist even in small host populations.
6 Control measures to combat wildlife diseases entail reducing either the total population size or the proportion susceptible to infection below some critical level.

Recommended supporting reading

The recent compilations by Grenfell and Dobson (1995), and Hudson et al. (2002) contain excellent chapters on the ecology of infectious diseases in wildlife populations, most notably the review of microparasite transmission and persistence by Swinton et al. (2002), and of the conservation implications by Cleaveland et al. (2002). Articles by McCallum, Barlow, and Hone (2001) and Lloyd-Smith et al. (2005b) provide critical assessments of the theoretical concepts incorporated into models. Somewhat different models are required for macroparasites like tapeworms and roundworms (Wilson et al. 2002). Indirect transmission via vectors like ticks or biting flies adds further complications (Randolph et al. 2001). Dobson (1995) outlines the use of a "who-acquires-infection-from whom" matrix to assess the transmission of rinderpest among multiple host species. The spreadsheet models developed for measles and HIV–AIDS were derived from lecture notes developed by C. Dye and B. Williams for a course at the London School of Hygiene and Tropical Medicine.

Exercises

1 Modify the measles model to represent cohorts (i.e., classes entering in the same year) moving through the school-going population. What fraction get infected during their first school year, and what proportion leave school without having had measles?

2 Establish the threshold population sizes for a range of diseases with different values for R_0.

3 Work out what modifications would be needed to the HIV–AIDS model to project a maximum prevalence of around 30% in the adult population, rather than the theoretical maximum of 85% for the most susceptible segment.

4 Following the outline of Barlow (1996), build a suite of models to assess the most effective control measures for wildlife diseases in different situations. Consider populations with density-dependent recruitment as well as density-dependent mortality, nonlinear forms of density dependence, and transmission rates saturating toward high host densities.

11

Scenario models

Exploring options

Chapter features

Topics

Forecasting; consumer–resource interactions; information; lag effects; dispersal sinks; thresholds; recovery rates

11.1 Introduction

In this chapter we confront a situation where a model could be an invaluable aid toward reaching a decision on the action needed to resolve some problem situation. The exercise may sometimes be structured formally as a workshop, with the modeler centrally involved in integrating the knowledge and understanding provided by the scientists, managers, and other stakeholders present. In other circumstances, one or more scientists might work privately on developing a model to aid their understanding of the situation, so as to give more informed advice. The problem could be, for example, how to enable the recovery of a fisheries resource that has crashed, how to reconcile forest management with preserving some endangered species, or how to manage competing interests of wildlife and livestock for vegetation and water resources in a region. Information might be quite sparse, and largely subjective, and the challenge is to synthesize what is known or believed, assess its reliability, and explore the implications of what it forecasts, given the uncertainties. Through modeling different scenarios, we investigate what might happen as a consequence of certain actions, taking into account various contingencies. There may be much guesswork, but at least the guesses will be informed by the best knowledge and opinion available at the time. Hence the decisions made will be defendable, even if things go awry later.

The scope of the model will generally need to be somewhat more complex than the population models we have developed thus far, so as to draw attention to wider consequences and ramifying effects. Economic aspects and other practical issues may also need to be incorporated, as you encountered in the hypothetical dodo example in Chapter 2.

The case study I will use is based on the wildlife management problem that confronted me during my doctoral research: how to respond to a developing overpopulation situation involving a previously endangered species, the white rhinoceros (*Ceratotherium simum*; Fig. 11.1). Although this example takes us back some time, the problem remains pertinent because seemingly overabundant herbivores continue to challenge wildlife professionals worldwide, notably deer in North America (McShea, Underwood, and Rappole 1997), and elephants in Africa (Owen-Smith et al., in press). Concerns relate to the impacts of the perceived overgrazing or overbrowsing on vegetation resources, the consequences for woodland regeneration,

Figure 11.1   A white rhino.

effects on other species being out-competed, losses in biodiversity, and the ethics of culling through killing. Also coming into consideration are dispersal movements, given brief attention in Chapter 9 when metapopulation models were introduced, as well as the restrictions imposed by fences or other barriers restricting dispersal. Lags in recovery and environmental variability are also additional considerations. Moreover, findings from the model helped bring about a change in the management policy, which is still being implemented today.

In this chapter, I will outline the background to the white rhino problem and the situation that had developed at the time of my doctoral study, as a context for the kind of model needed. The initial model was developed without the aid of a computer, and then taken further at a later stage. You will learn how an influential model can be developed despite meager information and much uncertainty about certain processes.

11.2   Background situation

The white rhinoceros had become almost extinct at the end of the nineteenth century. A remnant estimated to number only 20 animals persisted between the Black and the White iMfolozi rivers in the northern section of today's KwaZulu-Natal Province of South Africa. It was given protection through the establishment of the "Umfolozi" (the archaic colonial spelling) and Hluhluwe Game Reserves in 1897, shortly after the rinderpest pandemic had depressed wildlife numbers across the subcontinent. Thereafter the population grew steadily. Early ground counts during the 1930s recorded about 300 white rhinos. The first aerial count in 1948 revealed over 500 white rhinos whereas the 1965 census indicated that the population was approaching 1,000 animals. The conservation success became a source of concern, because by then the animals were having a notable impact on the grass cover in the reserve. Operations to move some of the animals to other areas were initiated. Around the same time the entire perimeter of the game reserve was fenced. Most of the rhinos removed initially were those outside the fence.

My study was initiated because of concern about the cause and consequences of the developing overpopulation situation. Grasslands in some sections had reached a state that would be regarded as "overgrazed" in a cattle ranch. Yet abundant grass remained in other areas hardly used by the rhinos. Hence my study was focused on factors influencing the distribution of the rhinos, such as territoriality. Varying opinions were held about the likely consequences if things were left alone. Some people projected a "rhino slum," with malnourished animals persisting in a degraded habitat, threatening the conservation of other species and the productivity of the ecosystem. They urged immediate intervention to reduce the white rhino population. Others maintained that the vegetation deterioration was merely a temporary response to drought, and that the vegetation would recover as it had done in the past (the "leave nature alone" stance). A major uncertainty was the effect that the fence would have on processes that might have operated effectively in the past when such barriers to movement did not exist.

Meanwhile, a census by helicopter conducted in 1970 indicated that the actual population numbered close to 2,000 white rhinos. This was 50% greater than the total counted the previous year by fixed-wing aircraft. Although initially received with disbelief by managers, this estimate was confirmed the following year when the helicopter count was repeated. Results showed a regional density exceeding 3 rhinos km^{-2} over the 450 km^2 area of iMfolozi Game Reserve, with local densities reaching 5 rhinos km^{-2} in some sections. Where was this population heading, and what might the consequences be?

11.3   Theoretical concepts

A belief that large herbivores introduced into a new environment, or released from predation, inevitably increase toward density levels at which they degrade the environment, and crash as a result, has formed a cornerstone of wildlife management. This perspective led to the influential model of herbivore–vegetation systems formulated by Caughley (1976a), building on the concepts outlined in Chapter 5 (see Figs 5.1 and 5.2). This model projects that oscillations can arise due to lags in the response of the vegetation to changing herbivore impacts, and by the herbivore population to the resultant change in food availability. Due to these lags, an overshoot of carrying capacity occurs, with resultant overgrazing and eventual stabilization of the herbivore population at a much lower abundance level as a consequence of the vegetation degradation (Fig. 11.2). Based on this model, Caughley (1976b) suggested that elephants and woodlands might cycle indefinitely without attaining any stable equilibrium, because of the long lag involved in the recovery of tree populations following damage by elephants.

Caughley's model was rather abstract and needed to be made somewhat more realistic before being applied to the white rhino situation,

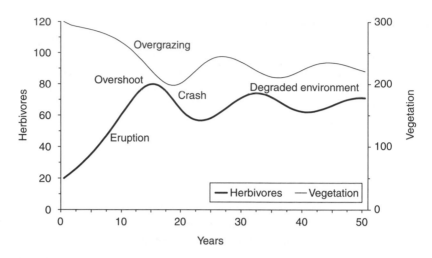

Figure 11.2   Output from Caughley's (1976a) model depicting the trajectory followed by an introduced herbivore population interacting with vegetation.

or indeed to elephants. Demographic lags could contribute to the over-shoot of carrying capacity; even if vital rates changed instantly to the values that would ultimately produce zero population growth, the size of the population would continue to expand for over a generation because of continuing recruitment into the reproductive segment of individuals already born. This demographic lag is one of the problems with regard to halting the growth of the human population. Furthermore, for large mammals density-dependent responses generally take hold at population levels quite close to the carrying capacity (Fowler 1981; McCullough 1992). The consequently steep "overcompensation" in the density feedback could also promote oscillations. A further consideration was research on elephants in East Africa suggesting that overpopulation situations developed where dispersal movements were blocked by fences or human settlements (Laws 1969). Recognition of the general importance of dispersal was reinforced by experimental studies on small mammals in North America (Krebs et al. 1973).

These are some of the concepts to be brought into a model of a white rhino population interacting with grassland resources. It would be useful at this stage to refresh your understanding of the consumer–resource models outlined in Chapter 5. This approach using coupled equations to represent the interacting populations will serve as the foundation for exploring the potential consequences of the grazing impacts of the white rhinos, as well as the wider ramifications for conservation objectives.

11.4   Modeling the white rhino–grassland system

11.4.1   Replicating the observed population trend

The first modeling need was to develop a sufficiently realistic model of a white rhino population and validate it against the observed increasing trend. The first step required was to establish the rate of population growth from the various censuses that had been conducted. The census totals needed to be corrected for rhinos removed to restock other areas, in order to reveal the intrinsic growth rate of the population, and correction factors needed to be applied for the varying accuracy of the different census methods. This analysis revealed that the total white rhino population had increased at a constant exponential rate of 9.5% per year through the decade 1960–72 (Fig. 11.3). It appeared that the population growth rate had been somewhat lower, about 6.5% per year, prior to 1960. Hence as the population grew larger, it increased faster. There was no indication of any reduction in the rate of growth despite the high density levels reached.

In order to build a demographically structured model, information was needed on basic reproductive and survival rates. I established that the mean intercalving interval was 2.5 years, with a shortest record, when the previous calf had survived, of just under 2 years. Females first gave birth around 7 years of age, and apparently carried on reproducing throughout their lives. Few adult deaths were recorded, indicating mortality rates of around 1.5% per year for adult females and 3% per year for adult males. Mortality losses among newborn calves amounted to only 5% over the first year. The study period was too short to estimate longevity. Cementum lines in the teeth of white rhinos that had died indicated a potential life span of around 40 years, although an Indian rhino held in a zoo had lived to 47 years. Accordingly, I assumed a maximum longevity of 45 years. Hence the age-structured demographic model needed to represent 45-year classes.

The model incorporating these vital rates projected an inherent population growth rate of 9.0% per year, after the age structure had stabilised. You can test this by running the program RHINODYN on the CD. The slightly higher growth rate of 9.5% estimated for the real population could have been because it had not yet attained a stable age distribution, having been increasing at a lower rate before 1960. The observed increase rate seemed to be close to the maximum intrinsic rate expected for such a large, long-lived mammal.

While setting up an equivalent spreadsheet model involving 45 year classes would be challenging, the white rhino population model can be

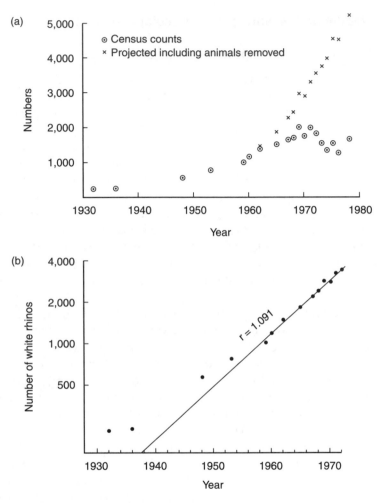

Figure 11.3 Pattern of growth of the white rhino population. (a) Recorded numbers within the park (circles) and projected total population including animals removed (crosses); (b) logarithmic plot showing overall exponential growth rate of 9.5% per year over 1960–72 (from Owen-Smith 1988).

approximated by aggregating the adult classes into one stage. Turn to Appendix 11.1 for guidance in setting up such a model, and establish how closely the population growth rate that it projects matches the observed population growth rate. Experiment by changing the vital rates slightly from those estimated above, and note how much difference this makes to the projected growth rate.

11.4.2   Assessing effects of the demographic lag

The next need was to assess the effect on the overshoot of carrying capacity of the delay in the age structure adjustment. In order to do this, the combination of vital rates that would confer zero population growth needed to be estimated. From studies on other megaherbivores (Laws, Parker, and Johnstone 1975; Hitchins and Anderson 1983), it seemed that most of the adjustment would take place in fecundity and juvenile survival. Through repeated trials, I established that the calving interval needed to be doubled, age at first parturition extended from 7 to 11 years (i.e., zero fecundity through the subadult age range) and mortality rates increased more than three-fold across all age classes, in order to transform the growing population into a static one (see Table 11.1).

While such changes in vital rates must eventually occur in response to rising population density, the abundance level at which the net population growth rate would become zero could only be guessed. I assumed that

Table 11.1 Formulation of density-dependent functions for annual mortality, natality, and emigration rates for the white rhino population model

Vital rate	Age range (years)	Value at maximum population growth	Value at zero population growth	Function (K = carrying capacity, H = herbivore biomass, number = years over which biomass was averaged)
First-year mortality	0–1	0.05	0.33	$\exp(-1.43 - 2.71\ln[K/H])$
Juvenile mortality	1–3	0.035	0.10	$\exp(-2.53 - 2.04\ln[K/H2])$
Immature mortality	3–6	0.025	0.073	$\exp(-2.8 - 1.7\ln[K/H3])$
Subadult mortality	6–10	0.015	0.05	$\exp(-3.2 - 1.71\ln[K/H3])$
Adult mortality	10–35	0.01	0.037	$\exp(-3.5 - 1.7\ln[K/H3])$
Subadult natality	6–10	0.45	0.00	$-0.75 + 0.80[K/H5]$
Adult natality	10+	0.40	0.22	$-0.05 + 0.30[K/H5]$

Source: From Owen-Smith (1988).
Note: The functions were designed to transform the vital rates observed in the expanding population into an appropriate combination for zero population growth.

"carrying capacity" could not lie much more than 50% above the prevailing biomass density, given the vegetation impacts that were occurring. I assumed further that the density feedback had a threshold onset, commencing at the current density. Specifically, both mortality and natality were assumed to be functions of the ratio K/H, with K representing the maximum density level and H the prevailing herbivore biomass. A power function was chosen so that mortality increased proportionately with proportional increases in the population biomass relative to the carrying capacity level. However, being long-lived animals with a reproductive cycle extending over several years, rhinos might not respond immediately to resource shortfalls. To allow for consequent lags in responses, the vital rates were made functions of the K/H ratio averaged over several preceding years. I assumed that, for juveniles, mortality losses depended on conditions over the preceding 2 years, while for adults both mortality and natality depended on conditions over the past 3 years. The fecundity of the subadult stage, effectively determining the age at first reproduction, was made a function of biomass over 5 years (Table 11.1).

Despite the age structure lags, this noninteractive model produced only a small (13%) overshoot of the equilibrium population level, with the population growth rate showing a ramp response to changing density (Fig. 11.4). This confirmed that, in order for wider oscillations to be generated, the model needed to incorporate the interaction between the white rhino population and vegetation. The age-structure effect alone was insufficient.

You should now modify the white rhino population model that you formulated to incorporate density dependence in the vital rates. Appendix 11.1 provides guidance on how to do this. Alternatively, explore the formulation written into the program RHINODYN.

11.4.3 Incorporating the grassland interaction

Caughley's (1976a) model assumed a logistic production function for the vegetation, with the plants regrowing continually to compensate for biomass removed by herbivores. In reality grasses grow when rains provide sufficient soil moisture, with little or no regrowth occurring during the African dry season. All of this dry grass can be consumed by herbivores, or by fire, without much effect on the amount of grass produced during the following wet season. Overgrazing is believed to occur when the grazing pressure during the growing season prevents grass tufts from storing sufficient root reserves to enable their recovery at the start of the following season, so that some tufts die. Hence the annual change in the grass population, in terms of the number of tufts and their size, needs to be

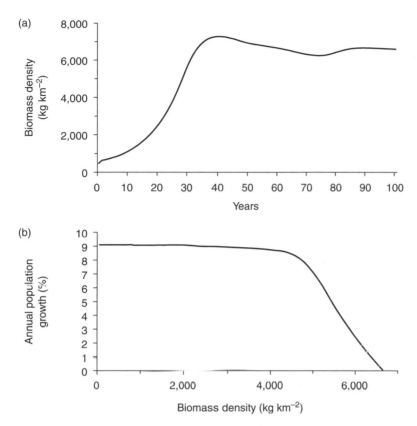

Figure 11.4 Output of the white rhino model showing small overshoot of the equilibrium population level due to age structure adjustments, in the absence of any vegetation interaction. (a) Growth of the white rhino population over time and (b) response of the population growth rate to changing population density.

distinguished from the seasonal production of the grass biomass providing food for white rhinos and other grazers by these tufts.

I assumed that the grass population, represented simply by a basal cover index, was reduced proportionately if consumption by white rhinos exceeded some threshold level. The threshold value was set at 50% of annual production based on agricultural guidelines. Specifically,

$$V_2 = V_1[1 - (IH - F_V/2)/F_V], \quad \text{if } IH > F_V/2, \qquad (11.1)$$

where V_1 is the initial grass cover, V_2 the grass cover following tuft mortality, F_V the forage biomass produced by the initial grass cover, I the annual consumption of grass per unit white rhino, and H the biomass density of the white rhinos. However, following tuft mortality space is opened for recolonization by new grass tufts, provided there is sufficient rainfall. Hence I assumed that some recovery of the grass cover could take place early in the next growing season, following a logistically saturating function. Hence the equation for the grass population dynamics is

$$dV/dt = r_V V(1 - V), \tag{11.2}$$

where r_V is the maximum proportional increase in the grass cover within a year. The vegetation carrying capacity is not represented because it has been assigned an indexed value of one, representing maximum cover.

For the annual consumption of grass by white rhinos over the seasonal cycle, I assumed a ratio-dependent hyperbolic function (see Section 5.4)

$$I = i_{max}(F_V/H)/\{f_{1/2} + (F_V/H)\}, \quad \text{if } (F_V/H) < 1, \tag{11.3}$$

where i_{max} is the maximum annual consumption per unit of white rhino biomass, (F_V/H) represents the food share available per white rhino, and $f_{1/2}$ represents the forage share at which the annual consumption reaches half of its maximum. I assumed further that the forage biomass F_V produced was proportional to the grass cover V, and in turn directly influenced the population level K of white rhinos that could be supported.

Based on a mean annual rainfall of around 700 mm, the annual grass production could amount to somewhere between 2,000 and 5,000 kg ha^{-1}. However, only a fraction of this would be available for consumption by white rhinos. Some grass would be inaccessible on steep slopes, some eaten by other herbivores, including insects as well as ungulates, and some of what remains too short to be grazed effectively. It seemed reasonable to assume that perhaps only around 25% of grass biomass produced would be effectively available. White rhinos consume about 1.5% of their body mass per day, expressed as forage dry mass, which extrapolates to an annual consumption amounting to about five times the live mass of white rhinos. I could not find any information on the inherent growth rate of a grass population. The only solution was to explore a range of possible values in the model.

The spreadsheet model of a white rhino population that you should have developed following the guidelines in Appendix 11.1 could now be extended to incorporate the vegetation interaction. This will enable you to explore the interactive dynamics of white rhinos and grasslands, which you could do alternatively by running the program RHINODYN. Note that the growth of the white rhino population includes a resource-dependent density feedback, in contrast to the exponential formulation of the numerical herbivore response in Caughley's (1976a) model. However, the amount of edible food produced by the vegetation, which is influenced by the grazing pressure of the white rhinos, determines the effective population biomass of white rhinos that can be supported. Hence overgrazing reducing the grass cover lowers the effective carrying capacity for the white rhino population.

If you run the model setting the saturation biomass of white rhinos at $10,000 \, \mathrm{kg \, km^{-2}}$ ($7.5 \, \mathrm{animals \, km^{-2}}$), you should find that the overgrazing threshold is exceeded only if the grass biomass available for consumption is less than $1,000 \, \mathrm{kg \, ha^{-1}}$. This follows from the fact that at most rhinos consume five times their own biomass over the course of a year (noting the conversion from square kilometers to hectares in the respective biomass units). Once overgrazing develops, the grass cover declines progressively causing the white rhino population to follow suit. The lags inherent in the coupled population responses generate oscillations with a period of around 40 years between peaks (Fig. 11.5). With either a low productive potential F_V, or low inherent recovery rate r_V, the grass population is depressed progressively toward its minimum level, and the white rhino population persists at greatly reduced abundance in this degraded environment (Fig. 11.5b). However, a grass with sufficiently rapid regrowth could support a huge white rhino biomass with little degradation (Fig. 11.5c).

11.4.4   Incorporating rainfall variability

In practice, the annual grass production will vary from year to year depending on rainfall. Hence the carrying capacity for white rhinos, which depends on the amount of forage produced, will also vary between years. However, the response of the white rhino population is dampened somewhat by its high inertia, that is, low annual mortality and slow growth rate. With a suitable choice of parameter values, the overgrazing threshold is exceeded periodically during drought years, even for vegetation parameters that would have been above this threshold under average conditions (Fig. 11.6). Explore the output patterns generated by different values for the growth rate and productive capacity of the grass layer, using either your spreadsheet model or the program RHINODYN.

These findings suggest how overgrazing might lead to progressive degradation of the grass cover and an eventual crash by the white rhino population, unless the recovery potential of the grasses is sufficiently rapid. Variable rainfall could result in thresholds being surpassed, precipitating periodic crashes in the herbivore population. The fundamental message is that the dynamics of white rhinos depend strongly on the dynamics of their resource base, as affected both by consumption and

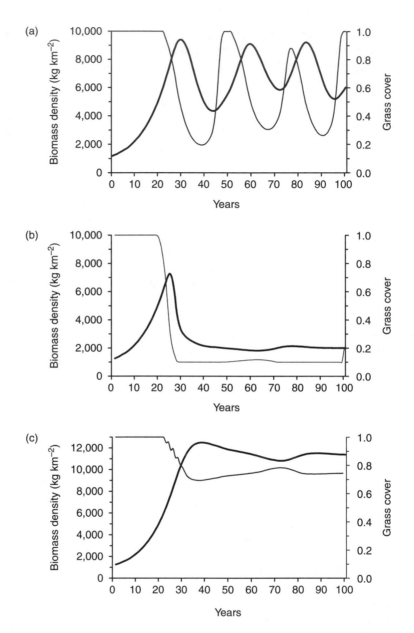

Figure 11.5 Output of the white rhino model for parameter settings causing the over-grazing threshold to be exceeded, under conditions of constant rainfall. (a) $r_V = 1$, $B_V = 600\,\text{kg}\,\text{ha}^{-1}$; (b) $r_V = 0.5$, $B_V = 500\,\text{kg}\,\text{ha}^{-1}$; and (c) $r_V = 2$, $B_V = 600\,\text{kg}\,\text{ha}^{-1}$.

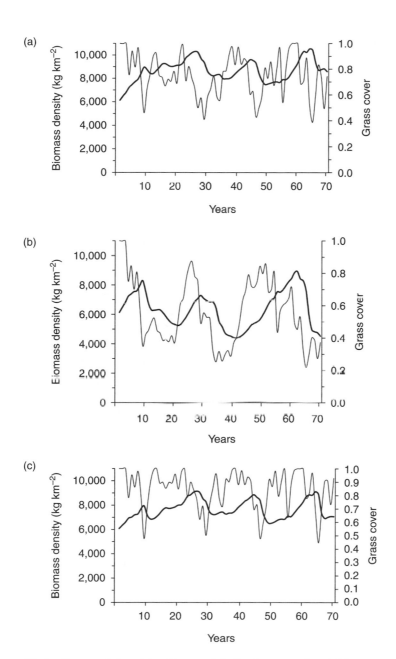

Figure 11.6 Output of the white rhino model allowing annual variability in rainfall. (a) $r_V = 1$, $B_V = 700\,\text{kg ha}^{-1}$; (b) $r_V = 0.5$, $B_V = 700\,\text{kg ha}^{-1}$; (c) parameter values as in (a), with dispersal allowed.

by rainfall. The uncertainty lies in whether the functional form, parameter settings, and threshold values assumed for the vegetation dynamics are realistic.

11.5 Exploring management options

The model illustrated the mechanisms that might lead to an overgrazing disaster. Given this threat, how should managers act? Should they be confident that the grass cover will always recover from periodic overgrazing during drought periods? Or should they intervene to ensure that a "rhino slum" does not develop, with its adverse consequences for other species and ecosystem processes as well as for white rhino conservation? And if they decide to intervene, what form should the intervention take?

An obvious response, widely adopted by wildlife managers, is preemptive culling, to keep the herbivore population below the level at which overgrazing develops. However, culling is costly and disruptive, even if undertaken by live capture and removal rather than by killing. Since the aim is not to obtain an economic yield from the population, the target should be to remove the minimum number of animals needed to keep the population below some acceptable abundance level. Which segments of the population, and in what proportions, should be taken out?

Young females that have just reached reproductive age make the greatest contribution to future population growth (from a theoretical perspective, they have the highest "reproductive value"). Older females have less life span left in which to make a contribution, while younger animals may not survive to reach reproductive age. Targeting subadults just before they produce their first offspring (to avoid dealing with young calves), 110 rhinos of both sexes would need to be removed annually from a total population of 2,000 animals to counteract the 9.5% population growth rate. If instead culling was unselective with regard to age and sex, it would be necessary to remove 174 rhinos per year.

However, maintaining the population of white rhinos at a ceiling of 2,000 is quite arbitrary. If the animals were allowed to experience nutritional deficiencies as a consequence of their vegetation impacts, the number needing to be removed could be reduced. The model suggests that the population could overshoot carrying capacity before these feedbacks become effective because demographic responses would be slow acting. Overpopulation situations might have been avoided in the past when animals could move elsewhere when food shortages started being felt.

There was evidence that dispersal had been occurring within the Hluhluwe-iMfolozi Park while the white rhino population grew. Specifically, the local rate of population increase had been lower in the core region where the density was highest than in peripheral areas, indicating a population redistribution outward from the core (Owen-Smith 1981, 1983, 1988). Local variation in the population structure showed that dispersal was primarily by subadults, of both sexes, plus some adult males seeking less contested areas to establish territories. Calculations indicated an emigration rate as high as 7.5% per subadult per year. How much difference could dispersal make to the white rhino–grassland interaction?

To investigate this, density-dependent functions for dispersal were added to the model. As was anticipated, dispersal movements dampened population oscillations, by keeping the population below the thresholds leading to overgrazing except during extreme drought years (Fig. 11.6c). The grass cover also remained generally in better condition. Moreover, the vital rates prevailing at the zero growth level appeared much healthier than for a situation where zero growth was attained in the absence of dispersal.

This scenario seemed more desirable, but where could the rhinos move, recognizing that the reserve was fenced? The solution was to establish dispersal sinks (or "vacuum zones") within the fenced confines of the park (Owen-Smith 1981). This entails concentrating removals in these designated regions. Animals running short of food within the population core have the opportunity to move into the sink zones. In effect, the animals decide when the population has exceeded carrying capacity by voting with their feet. The managers merely help them over the fence, and on to distant, less crowded pastures on the back of a truck. This sink management achieves two ends. It allows the rhinos to move out of crowded areas before overgrazing thresholds are passed, and the lightly grazed sink areas provide a habitat refuge to animal or plant species adversely affected by the grazing impacts of the white rhinos.

Dispersal sink management has become integral to the "process-driven" management adopted for the Hluhluwe-iMfolozi Park. Rhino removals continue to take place annually, effectively keeping the population below the peak abundance attained during my study, but are restricted to the designated sink zones. Rhinos elsewhere in the park experience no disruption of social processes. Animals have been captured alive and transported to other protected areas, private wildlife ranches, and some zoos. As a result the total number of white rhinos has increased to over 11,000, widely distributed across southern Africa, so that the species is no longer on the endangered species list.

11.6 Overview

The model focused attention on processes governing the interaction between a herbivore population and its food resource base. It highlighted the importance of rates of resource regeneration, critical thresholds, the effects of rainfall variability, and the importance of dispersal movements as a short-acting response to deteriorating conditions. Yet the model itself was relatively crude. Vital rates for the herbivore population were estimated from a brief study and showed no evidence of density feedbacks. The ceiling abundance level was a pure guess. The grass resource was considered simply as one aggregate population, with no consideration of the grass species constituting the herbaceous cover. The recovery rate of the grass population following overgrazing was just a guess, and the fraction of grass production estimated to be available for consumption by white rhinos was quite arbitrary. It is questionable whether the simple logistic model provides an adequate description of grass dynamics. Overall, the model cannot be defended as an accurate representation of the white rhino–grassland interaction.

Nevertheless, the modeling exercise succeeded in highlighting the importance of dispersal, and thereby led to a change in management. The model itself has not been used as a management tool. However, this early experience in modeling made me aware just how powerful this tool can be as a learning aid. From this exercise, you should have learned the following:

1 Much can be learned from a model, even when the available information to construct it is limited.
2 The model helps focus attention on possible future outcomes.
3 The most crucial factors influencing the outcome can be established, indicating where further research is most needed to resolve uncertainties.
4 The consequences of alternative management responses can be explored, helping ensure that actions are effective in achieving their aims.
5 Modeling can be influential in leading to a change in a management policy.
6 For interactive systems where consumers have a substantial impact on their resource base, the consumer dynamics depend greatly on the regeneration dynamics of resources.
7 Constraints on movements can amplify the potential for oscillations and hence the risk of biodiversity losses.

Recommended supporting reading

Chapters in McShea et al. (1997) describe some of the problems that have arisen with regard to overpopulations of deer. The development of the white rhino model is explained in more detail by Owen-Smith (1988). A conceptual framework for accommodating multiple food types and seasonal vegetation dynamics is developed in Owen-Smith (2002a). Some of the dilemmas that have arisen with regard to the need for culling to manage the burgeoning elephant populations in many protected areas in southern Africa are discussed by Whyte, Van Aarde, and Pimm (1998), Van Aarde et al. (in press), Gillson and Lindsay (2003), and Owen-Smith et al. (in press). Owen-Smith (2002b) outlines how modeling indicates that spatial heterogeneity in the environment might stabilize the interaction between elephants and trees.

Program on the accompanying CD

RHINODYN

Exercises

1 Replace the logistic growth model for vegetation dynamics in the rhino–grassland model with a constant production rate independent of the vegetation biomass, presumably coming from large underground reserves. Establish what difference this change makes to the dynamics of the herbivore–vegetation interaction.
2 Alter the parameters governing herbivore population dynamics to represent a medium-sized grazer like a wildebeest. Assess what additional influence predation on the herbivore population could have on the interactive dynamics.
3 Alter the vegetation model to represent woody plants responding to deer browsing. Determine under what conditions overbrowsing could lead to the collapse of the deer population, and the effect of the deer overpopulation on the subsequent recovery of the trees.
4 Develop a spatially compartmentalized model to represent dispersal movements redistributing the herbivore population between regions differing in vegetation attributes. Use this model to suggest where dispersal sinks should be located to be most effective.

12

Vegetation models

Biomass to gap dynamics

Chapter features

Topics

Seasonal production and attrition; biomass components; ingrowth; growth suppression; size structure; contingent transitions; seed banks; episodic recruitment; even-aged cohorts; disturbances; gap colonization; structural transitions

12.1 Introduction

Plants serve both as a resource providing timber, fruits, material for crafts and medicines, for direct human use, and as a food resource maintaining herbivore populations. Their population dynamics differ from those of animal populations in some fundamental ways. Counting individuals is problematic because what constitutes an individual is difficult to decide, hence some other measure of abundance is needed. Plant parts die and are lost independently of the survival of the plant, leading to seasonal production and shedding of leaves and supporting tissues, besides changes in the population. Most of the above-ground tissues of grasses can be destroyed by fire, yet restored completely the next growing season. Tissues consumed by herbivores can be replaced through vegetative growth within the same season. Seeds in the soil form a cryptic segment of the population, commonly overlooked. Transitions from seed to seedling, and seedling to established plant, can be contingent on the prevailing conditions, rather than proceeding automatically each year. Establishment depends critically on the availability of space, in particular gaps opened by the death or shrinkage of other plants. Disturbances of various kinds play an important role in opening gaps. Growth between stages is indeterminate, and plants may be held as seeds or seedlings for long periods until conditions permit growth toward reproductive maturity. Access to sunlight depends strongly on height, so that prior establishment confers a competitive advantage. Plants are fixed in space, so that competitive interactions occur mainly with neighbors.

Which form of model is most appropriate depends on the purpose for which the model is intended, more especially for plants than for animal populations. If our concern is simply with vegetation as food for herbivores, a model considering only the aggregate dynamics of available plant biomass might be adequate. If plants represent resources in their own right, more attention needs to be given to processes governing transitions through life history stages. Since established plants occupy space, it can be more insightful to focus on the dynamics of gap formation and occupation than enumerating numerous seedlings, most of which make no contribution to future populations.

In this chapter, I will outline three distinct approaches toward modeling vegetation dynamics:

1 seasonal biomass dynamics of a forage resource supporting herbivores;
2 stage matrix representation of trees potentially harvested for timber; and
3 patch occupation dynamics in the context of fire, climatic extremes, or other disturbances.

12.2 Seasonal biomass dynamics of vegetation supporting herbivores

12.2.1 Basic model structure

The consumer–resource models developed in Chapter 5 assumed that the vegetation resources supporting a herbivore population regenerate continuously in response to consumption by herbivores, over an annual time frame. In reality, plants undergo a seasonal cycle of growth and attrition, which has important consequences for herbivore population dynamics. Grasses regenerate leaves and stems at the start of each growing season, from tissues persisting underground. Woody plants likewise commonly

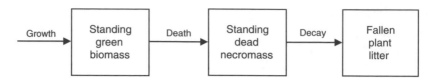

Figure 12.1   Conceptual model of grass biomass dynamics between standing green, standing brown, and fallen litter components.

grow new leaves each spring. When summer or the rainy season ends, grass parts above-ground progressively become dead and moribund, and hence of reduced nutritional value. Many trees shed their leaves in autumn, or part-way through the dry season. Green forage becomes a nonrenewing resource, reduced further by grazing or browsing. The winter or dry season bottleneck in food quantity and quality places the major limitation on the growth and survival of herbivores. Moreover, the forage resource is generally constituted by several plant species, differing in their growth dynamics and nutritional value. It matters little whether this forage resource is restored at the start of the new growing season through the appearance of new plants, or via the growth of established plants.

The total grass biomass above-ground can be partitioned between green material V_g (biomass in the strict sense) and standing dead tissues V_d (brown "necromass"; Fig. 12.1). Since leaves have a limited life span, there is ongoing turnover from green leaves to brown leaves during the course of the growing season. The brown material eventually falls to the ground as litter and decays. When growth ceases during the winter or dry season, there is a rapid conversion of remaining green material to brown. This can be represented by a logistic growth model using weekly (or even daily) time steps, with the saturation biomass level dependent on the combined shading effect of both green and brown leaves (see Appendix 12.1). A very simple seasonal cycle can be used, with growth occurring continuously during a 26-week growing season, then ceasing during a 26-week dormant season, which seems most appropriate for African savannas and other drylands where grass growth is controlled primarily by rainfall (Fig. 12.2a). A sinusoidal growth pattern might be more appropriate for temperate climates where temperature is the main controlling influence. Regrowth is initiated at the start of a new growing season from some minimum biomass translocated from stored reserves. Remaining dead material may hamper this regrowth. However, in the model it has been assumed that this top-hamper decays more rapidly during the wet season when decomposer organisms are more active.

Herbivores consume a combination of green and dead leaves, but with a preference for the more nutritious green leaf component. The material

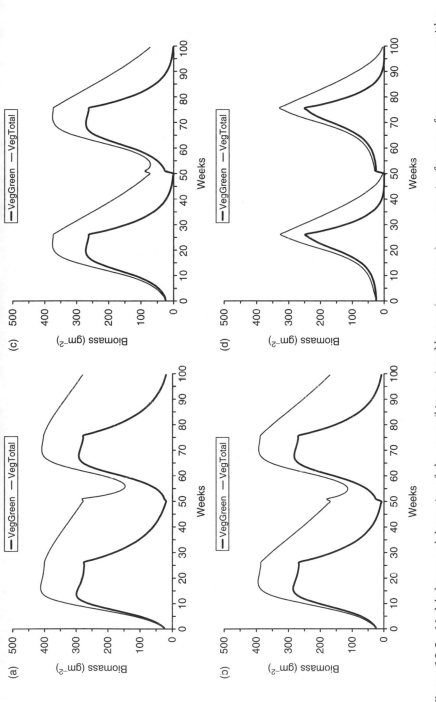

Figure 12.2 Modeled seasonal dynamics of the green (biomass) and brown (necromass) components of a grass forage resource with a regular alternation between a 26-week growing season and 26-week dormant season, for fixed stocking levels of herbivores. (a) Ungrazed; (b) low stocking level, (c) moderate stocking level and (d) high stocking level.

they remove during the growing season has only a minor impact on the grass biomass, because it can be partly replaced by compensatory regrowth. During the dormant season the grass biomass gets progressively reduced through grazing as well as intrinsic decay, so that with a high herbivore stocking level very little vegetation may remain by the end (Fig. 12.2b–d). The quality of the forage consumed also declines as brown tissues predominate, restricting the productive capacity of the herbivores. The herbivore biomass that can be supported depends on the counterbalancing between the growth potential enabled during the wet or summer period when food is abundant and of a high quality, and the attrition in this animal biomass over the dry or winter season when the quality and perhaps also the quantity of this food is low. At some stocking level, so little food remains toward the end of the adverse season that the net growth in herbivore biomass becomes zero.

In reality, rain does not fall during every week of the growing season, and some rainfall may occur even during the dry season weeks. If the soil is wet, grass grows, otherwise growth is checked. This means that grass growth occurs somewhat erratically, affecting the extent to which forage is grazed down (Fig. 12.3). This pattern can be introduced into the model by assigning a probability level to the occurrence of rainfall, then using a random number generator to decide whether rain does actually fall and promote grass growth in a particular week, as described in Appendix 12.1 The impact of grazing on grass growth, and hence on the herbivore biomass that can be supported, is particularly sensitive to any delay in the commencement of grass regrowth at the start of the growing season, and to rain generating pulses of green regrowth during the dormant season. Furthermore, a favorable year can elevate the herbivore biomass above the sustainable stocking level, leading to greater impacts on the grass layer the next year.

12.2.2   Assessing how grassland composition affects grazing capacity

In reality, a grassland consists of several grass species differing in their potential growth rates and in the forage biomass that they attain. One of the issues in range management is how the composition of the grass layer determines the area of vegetation needed to support each herbivore. Grass species are rated in terms of their forage potential, and together with their contribution to the grass cover, this determines a range condition score. Grassland with a higher score is expected to support more herbivores. However, if the herbivore stocking level is set too high, the better grass species

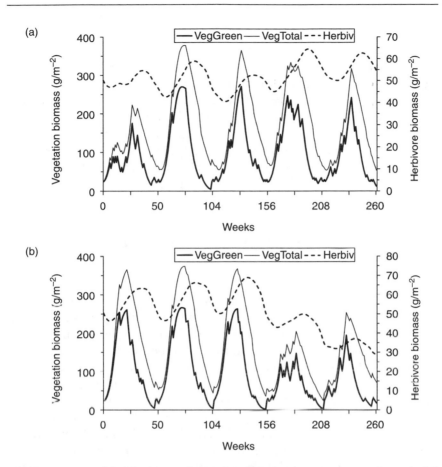

Figure 12.3 Modeled dynamics of grass forage resource in a situation where rainfall varies randomly between weeks, with a 0.8 weekly probability during the wet season and 0.2 during the dry season. Panels (a) and (b) represent outputs of two trial runs both with moderate stocking density of herbivores responding to food availability. In (a) the herbivore population is sustained, while in (b) it crashes through over-grazing.

decrease in their representation as a result of the higher grazing pressure that they receive, and the grassland score deteriorates. Nevertheless, some of the grass types that increase under heavy grazing are lawn-forming grasses of high nutritional value. The limitation they impose is that they produce less grass biomass for carryover into the dry season. These lawn grasses are favored by white rhinos as long as they retain sufficient biomass (see Chapter 11). Is it indeed true that grassland with a prevalence of lawn-forming species has a lesser capacity to support herbivores than grassland in which the "decreaser" species predominate?

The program RANGECOND on the CD was written to assess the relation between grassland composition and the productive potential of the herbivores supported by this grassland (borrowed from chapter 11 of Owen-Smith 2002b). It requires a food choice algorithm determining how the herbivores select their diet from among these grass types as their availability and quality changes over the seasonal cycle, and hence is not easily replicated in spreadsheet format. It also presents the metabolic balance of the herbivores in some detail. How to incorporate adaptive behavior into a model will be addressed in Chapter 14. Nevertheless, you may want to try running this model at this stage to explore the output that it generates.

12.3   Size-structured dynamics of a tree population

12.3.1   Basic tree model

The growth and survival rates and seed production of trees depends strongly on size, making a size-structured matrix the appropriate model to use. The size classes should be constituted so that trees can potentially grow into the adjacent size class during the time step in the model. For a tree potentially living several hundred years, a time step of a decade would be appropriate. The best measure of size seems to be stem diameter. This assumes that the time required for a small tree to grow from 10 to 20 cm in diameter is similar to that required for a big tree to grow from 100 to 110 cm in diameter. Growth actually entails the addition of an annulus of bark and outer woody tissues to support the extra height and foliage of a larger tree, while the inner wood is largely dead necromass. Of course, some trees may grow less fast than this, and hence remain in the same stem diameter class at the end of the time step.

Trees can produce enormous numbers of seeds, suggesting that a tree population could increase hugely between one time step and the next. However, counting the number of seeds or tiny seedlings is not an appropriate measure of the size of a tree population. Biomass, or something related to it, would be a better measure. Moreover, most of the seeds produced come to naught in terms of augmenting the population, being consumed by the numerous animals that depend on seeds for foods. Indeed, trees respond to the high seed predation by patterns such as mast fruiting, that is, years when vastly greater than usual numbers of seeds are produced, so that some have a chance of escaping the granivores that may consume almost every seed in most years. In practice, it is almost impossible to count the number of seeds entering the seed bank in the soil, and difficult enough to determine the density of small seedlings emerging from these seeds.

Hence the top line of the transition matrix commonly represents merely the "ingrowth" into the smallest size class that can be enumerated.

The final consideration is how density, or "crowding," affects the dynamics of a tree population. Botanists recognize a self-thinning rule whereby plants cannot grow larger until some of the existing plants have died, opening space into which neighboring plants can grow. This has been documented both for planted crops, and for tree plantations, which obviously can fill all of the space that is available after reaching a certain size. Once this stage is reached, the population declines numerically while individual plants grow larger, without much change in the total standing crop or leaf area. If no plants die, growth becomes stunted. This suggests that the density feedback affects largely growth between size classes, rather than survival within a size class. Nevertheless, through being held in a smaller class plants may experience a higher risk of death than would have been the case had they grown larger.

These concepts have been incorporated into the True BASIC model TREE-DYN on the CD. Density dependence can be switched off to enable the intrinsic growth rate generated by a particular combination of vital rates to be determined. The default subroutine for density dependence slows growth between size classes as the population biomass (in arbitrary units) approaches some maximum limit. The output plots the size distribution of the population as well as offering optional graphs depicting the population dynamics and form of the density response. You should find that the size distribution approximates an "inverse-J" in form, with a high number of the smallest size class and progressively lower numbers of larger trees (Fig. 12.4).

Using this model, you can explore how size does not necessarily represent age in a tree population. To do so, initiate the population with a cohort of seedlings, and observe how these seedlings later become distributed over several size classes. This results from the suppression of growth as crowding relative to the biomass that can be supported develops. Small saplings can remain in the understory of forests for many decades, awaiting a canopy opening to enable them to grow taller.

12.3.2   Tree harvesting model

A very simple model developed to aid the harvesting of tropical hardwood trees in Nigeria was described by Osho (1991, 1995), together with measurements on changes in the size structure of a sample stand over an 18-year period. These enabled the proportion of trees surviving as well as the proportion that had grown into a larger stem diameter class to be estimated, providing the basis for a matrix model (Table 12.1).

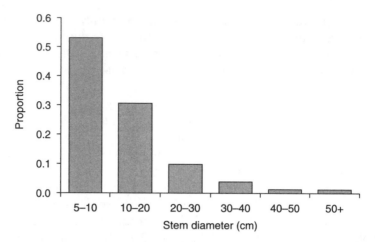

Figure 12.4 Inverse-J distribution of size classes of trees observed in a tropical rainforest in Nigeria (data from Osho 1991).

Table 12.1 Parameter estimates for an indigenous tropical forest stand in Nigeria

	Size class					
	1	2	3	4	5	6
Stem diameter range (cm)	4.8–10	10–20	20–30	30–40	40–50	50+
Ingrowth contribution per tree		0	0.73	2.24	5.04	7.42
Proportion persisting in class	0.42	0.62	0.66	0.79	0.62	0.89
Proportion growing into the next class	0.06	0.05	0.14	0.20	0.2	
Overall survival of class	0.48	0.67	0.80	0.98	0.82	0.89
Initial number of stems in 1.5 ha	980	599	193	63	28	19

Note: Size classes represent constant stem diameter increments, rates are relative to an 18-year time period (from Osho 1991, 1995).

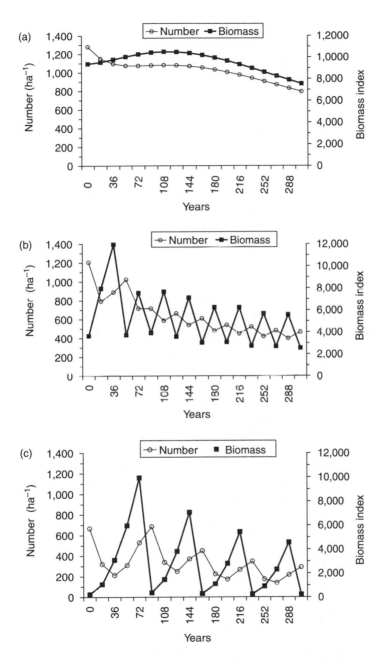

Figure 12.5 Projected growth in numbers and biomass of a tropical tree stand in Nigeria (a) based on observed vital rates, (b) with harvest removing all trees in size classes 4–6 every 36 years, and (c) with clear-felling of all trees but the smallest class every 72 years.

This model can be represented easily in a spreadsheet and used to explore the consequences of different harvesting policies (see Appendix 12.2). The first need is an estimate of the inherent growth rate projected by the vital rate estimates. For timber production we want to know not just the numerical increase in the number of trees, mostly in the smaller size classes, but rather the growth in total weight of potentially harvestable wood. This can be assumed proportional to tree volume, which will be roughly a cube function of the median stem diameter in each class. The observed data project a steady decline in the number of trees, while the estimated timber biomass rises at first and then falls (Fig. 12.5a). The population change, measured either as timber mass or numerically, eventually stabilizes at a 4.7% decline per 18 years. The negative growth potential probably reflects the fact that the stand had not previously been harvested, apart from selective logging of one highly valuable species, so that space was saturated.

The maximum growth potential of the stand can be estimated by assuming that survival rates remain unaltered, with all surviving trees growing into the next larger class within the 18-year time step. This yields a 40% increase in biomass over 18 years, after the size structure has stabilized, which is equivalent to a 2% annual growth rate. A higher growth rate could result if tree fecundity was also enhanced in the absence of crowding, represented by ingrowth of plants into the smallest stem diameter class.

Some of this growth potential could be released by harvesting some trees from the stand. The model thus needs to represent how growth between size classes is enhanced if the overall stand biomass is reduced to a certain extent. We will assume that the overall survival rate for each size remains unaltered, while the proportion moving into the next size class increases. The consequences of alternative harvesting policies then need to be assessed, in terms of the size classes to be removed, and the interval between harvests. One option is clear-felling, leaving only the smallest stems to regenerate the forest. Another option is the selective removal of trees in the largest size class, which have little further growth potential. Various intermediate policies could also be adopted. For each policy, the appropriate harvest interval needs to be identified.

The model output suggests that the highest yield under each policy would be obtained by (i) removing trees in the largest size class every 18 years, (ii) removing all marketable trees above 30 cm stem diameter every 36 years (Fig. 12.5b), or (iii) felling all but the smallest class every 72 years (Fig. 12.5c). The time step in the model does not allow a finer resolution of the optimal interval. Modeling indicates further that the second policy gives the highest yield over a 300-year period. This entails felling trees before they reach the largest class, with the lowest growth potential.

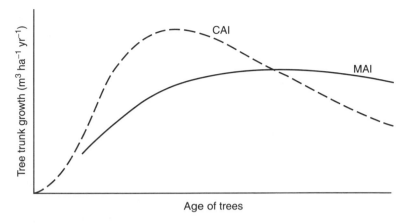

Figure 12.6   Changes in annual stand increment in tree biomass and in mean annual increment over the period since clear-felling with time since establishment of a cohort of trees (from Newman 1993).

Clear-felling all but the smallest stems gives the poorest return because the recruitment potential of the trees removed is also lost. Osho (1995) reported somewhat different findings, but he did not allow for compensatory growth by the remaining trees after harvests. Selective removal of only a portion of the trees in particular size classes would be advantageous for forest conservation, but would give a lower return in terms of timber produced.

Newman (1993) outlines some of the principles used in deciding on the optimal rotation period for plantation forestry. The timber yield is maximized by felling at the stage when the mean annual increase in biomass since previous removals is greatest, which tends to be slightly later than when the annual wood production peaks (Fig. 12.6).

12.3.3   Woodland dynamics with episodic recruitment

In woodlands growing in drier climates, seedling establishment may be successful only at irregular intervals. Sufficient rain at the right time for seed germination needs to be followed by enough rain to allow these seedlings to extend their roots sufficiently to become established. The abundance of granivorous and browsing animals must also be low enough to allow some seeds and seedlings to escape being eaten. This leads to episodic pulses in recruitment. By producing lots of seeds, trees gamble on the occurrence of favorable conditions. Each tree needs to recruit only one or two offspring to the mature stage (depending on whether it is monoecious or dioecious) during a life span of many centuries to maintain a population. Recruitment events may take place only 2–3 times within a century (Wiegand, Jeltsch, and Ward 2004).

The program TREEPREC incorporates episodic recruitment from a persistent seed bank, using a contingent rule to decide whether the transition from seeds to established seedlings occurs during any time step. Conditions during each decade are defined simply as "good" or "not good" for successful establishment, with "good conditions" occurring randomly with some specified probability each decade. With episodic recruitment, the size structure distribution of the tree population can vary quite widely, at any particular time, depending on when successful recruitment last occurred. Stands of mature trees with few small plants may develop, giving a direct-J distribution of size classes (Fig. 12.7b), or the population can show a humped structure with the majority of plants in intermediate size classes (Fig. 12.7c). Explore how these patterns develop using this model, both varying the recruitment interval and undertaking repeated runs with the same recruitment interval. Since the model is stochastic, you will find different patterns at the end of each run. With episodic recruitment, no stable size distribution is attained, and the population trend over time will reflect the recruitment pulses.

12.4   Gap dynamics model

An alternative modeling approach for forest dynamics is to consider not the number of plants in the population, but rather whether sites that could support a tree are occupied or not, and the size of the dominant plant present. The number of plants occurring at a site is irrelevant because only the largest of these will grow to maturity, shading out the others. Besides the eventual death of mature trees, disturbances of various kinds can open gaps for colonization. Successful occupation of these sites depends on processes governing the dispersal of seeds, or release from shading of seedlings that had previously established. Recruitment is constrained by the availability of gaps, rather than by the production of seeds or seedlings. Hence these models emphasize the interplay between disturbances and the conditions enabling seedling establishment, both of which occur at irregular intervals. Brown and Wu (2005) documented how climatic variability interacting with fires resulted in recruitment pulses at intervals of a century or longer in ponderosa pine woodlands in Colorado.

The consequences of disturbances interacting with recruitment pulses can be explored using the program TREEGAP. This model represents the

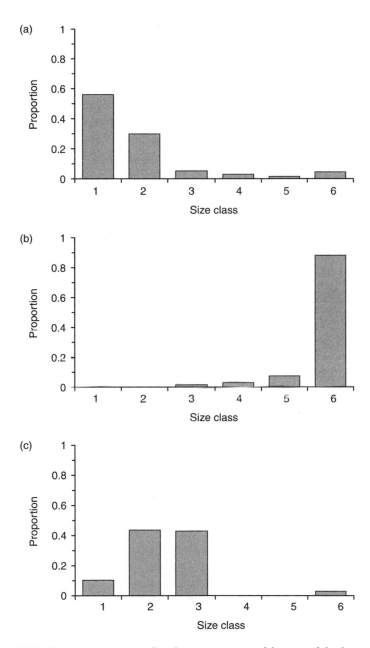

Figure 12.7 Tree size structure distributions generated by a model when recruitment occurs irregularly, (a) inverse-J distribution with predominance of small trees, (b) J-distribution with predominance of big trees, and (c) humped distribution with predominance of medium-sized trees.

proportion of sites that could potentially support a large tree that are occupied by trees of a particular size, ignoring smaller plants within each site that will be outcompeted. Gaps exist if the total number of trees is less than the total number of sites. If you run this model with regular recruitment every decade, and no disturbances, the majority of sites becomes occupied by trees in the largest size class, with no vacant gaps (Fig. 12.8a). Gaps arising through the death of a tree are immediately occupied by seedlings. This clearly represents a forest situation. The forest becomes even more densely thronged with canopy trees if you increase the tree survival rate per decade, effectively lengthening the longevity of the largest size class.

By allowing disturbances causing enhanced tree mortality at intervals of around 50–100 years, the forest becomes transformed into a woodland with a more mixed size structure and some gaps (Fig. 12.8b). The woodland can be transformed into an open savanna by making disturbances frequent and the interval between successful recruitment events relatively long. Higgins, Bond, and Trollope (2000) developed a model of savanna tree dynamics based on the interaction between frequent fires, which kill a proportion of trees and transform others to smaller size classes, and episodic establishment. Since savanna trees are shorter-lived than those typical of forests, a time step of 5 years may be more appropriate than a decade. This approximates the fire return interval typical of savanna regions. If the disturbance caused by fire occurs regularly every 5 years, but successful recruitment only once every 15–20 years, the woodland takes on the aspect of an open savanna, with a large proportion of sites not filled by trees or even seedlings (Fig. 12.8c). For the default parameter values in the TREEGAP model, a critical transition in the savanna structure takes place between a recruitment interval of 10–15 years and one of 20–25 years. You will need to undertake repeated trials to confirm this, because of the stochastic nature of the recruitment events. Lengthen the interval between fire disturbances, and observe what effect this has on the savanna form. Longer fire intervals could result if heavy grazing pressure reduces the fuel load for fires, except in years of high rainfall.

The model also allows the growth rate of trees between size classes to be slowed down, which might happen if temperatures were cooler, representing the conditions that prevailed during the last ice age around 20,000 years ago. If you slow down the growth rate sufficiently, allow fires to occur regularly within every 5 year period, and allow a sufficiently long interval between recruitment events, the savanna can become transformed into an open grassland. This may take several centuries, because while some trees remain they continue to supply seeds to recolonize gaps. After the mature tree class drops below some threshold level, their seedling

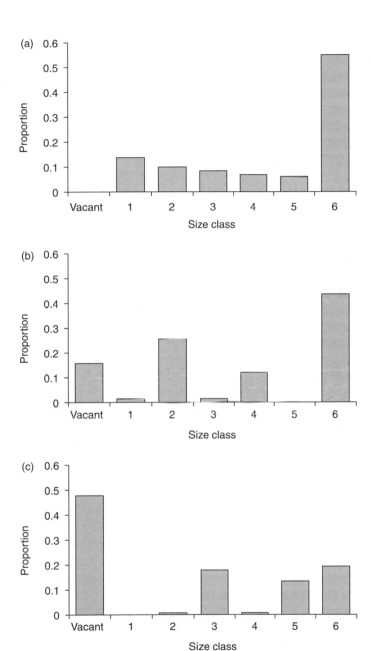

Figure 12.8 Output of gap colonization model showing the size structure of trees dominating sites as well as proportion of sites not occupied by trees (a) with regular recruitment and infrequent disturbances, (b) with infrequent disturbances and irregular recruitment, and (c) with frequent disturbances and irregular recruitment.

input becomes inadequate, and trees disappear from the landscape. The dispersal rates of tree seeds into gaps can also be reduced in the model, as might happen when the birds or mammals responsible for such dispersal start to become rare because of the habitat change.

Currently there is greater concern with the consequences of the carbon fertilization associated with global warming, which speeds the growth of woody plants. Increase the growth rate factor in the TREEGAP model, and note how the savanna structure is altered as a result.

12.5 Overview

This chapter provided you with various alternative models for representing the dynamics of plant populations, from changes in biomass components through size-structured matrices to gap dynamics. They represented a range of vegetation structures from grassland through forest, woodland, savanna, and back to open grassland. Models may be applied to project forage production for herbivores, timber production for humans, or merely the vegetation physiognomy serving as a habitat for other organisms.

From this exercise, you should have learned the following:

1 Plant population dynamics can be represented in different ways depending on the application of the model.
2 Forage production models require subannual time steps to represent the seasonal fluctuation in forage quantity and quality limiting herbivore population growth.
3 Tree management for timber production depends on the processes governing the growth of plants between size classes.
4 Episodic or irregular recruitment can produce size class distributions that deviate from the inverse-J distribution expected if recruitment occurs regularly.
5 Even-aged cohorts may be derived from contingent events influencing seedling germination and establishment.
6 Gap dynamics models represent the interactions between disturbances, recruitment pulses, and dispersal processes.
7 A forest can be converted into a savanna woodland by increasing the frequency of disturbances while reducing the frequency of successful seedling recruitment.
8 Cooler temperatures slowing growth between tree size classes could transform a savanna into an open grassland, if disturbances are sufficiently frequent relative to recruitment events.

Recommended supporting reading

The book by Silvertown and Doust (1993) gives an excellent overview of plant population processes, especially Chapter 6. Alvarez-Buylla (1994) describes the application of matrix models to forest tree dynamics, at stand, patch, and gap scales, in some detail, including density-dependent influences. Peters (2001) reviews ways in which plants are exploited by humans for a variety of products. Baxter and Getz (2005) present a detailed model of tree dynamics in the context of impacts from elephants as well as fire and climate.

Programs on the accompanying CD

RANGECOND; TREEDYN; TREEPREC; TREEGAP

Exercises

1 Develop a model of the biomass dynamics of woody plant foliage serving as a forage resource for deer, or any similar browser.
2 Evaluate the conclusions reached by Osho (1995) for evaluating optimal harvesting policies for tropical rainforests in Nigeria, making different assumptions about how compensatory growth occurs.
3 Use the tree dynamics program, or your own equivalent, to investigate how the size distribution of a tree population is influenced by patterns of recruitment.
4 Establish the relationship between potential tree longevity (the inverse of the survival rate of the largest age class, plus the time taken to reach this size class), and the threshold interval between recruitment events, enabling a tree population to persist.
5 Investigate what changes in the frequency of recruitment events relative to disturbances are necessary to convert a forest into an open savanna woodland, and how much reduction in the growth rate of the trees would be needed to produce a treeless grassland.
6 Compare the output of the simple gap dynamics model of savanna woodland with that projected by the more mechanistic grid-based model developed by Higgins et al. (2000). What are the restrictions of the simpler model and the benefits of the more elaborate model?
7 Explore the consequences of CO_2 fertilization and global warming, enhancing the growth rates of trees between size classes, on the structure of different vegetation types.

13

State transition models

Habitat patch dynamics

Chapter features

Topics

Patch states; Markov matrices; vegetation succession; transition rates; river channel dynamics; episodic disturbances; stochastic processes; savanna dynamics; herbivore stocking strategies; grid-based models; neighborhood effects

13.1   Introduction

The models to be covered in this chapter represent changes in the state of an assemblage of plant species determining habitat conditions. These states, defined by the kinds of plants and their structural physiognomy, are distributed in recognizable patches across some regional landscape. External influences, management actions, and time can change the relative proportions and juxtaposition of the habitat types represented. In Chapter 12, we considered how episodic disturbances, coupled with intermittent recruitment, could result in the transformation of a wooded forest or savanna into an open grassland. We now consider the mosaic interspersion of patches representing forest, savanna or grassland, or some other states across a wider landscape.

The modeling approach addressing changes in habitat states over time uses Markov transition matrices. The key difference from a population projection matrix is that the transition probabilities from any state must now add up to 1.0. Population matrices omit the transition from living to dead because animals or plants that have died are not considered part of the population. In a Markov model, the mortality rate would need to be included, although death is an "absorbing state" from which there is no return. Hence Markov models produce no inherent growth rate, and interest lies in changes in the proportion of the environment constituted by different habitat types.

The first step in formulating a Markov model is to identify the functionally distinct habitat types, for example, the distinction between a grassy meadow and a grove of trees. The probability that the meadow could be invaded by trees to become a shrubland, or the woodland burned and replaced by grassland, at some later time needs to be considered next. The transition probabilities are derived from long-term monitoring of changes in vegetation states in some region, or perhaps from aerial photographs. The appropriate time step spans several decades or even centuries. The Markov model could be deterministic, projecting only the average distribution of states to be expected, or stochastic, identifying all the possible state distributions that could arise, depending on chance events, and their relative likelihood. Strictly, we are dealing with "first order" Markov chains, which means assuming that the transition probability depends only on the current state of a patch, and not on how this state was reached.

Models projecting merely the state proportions can be taken a step further to represent explicitly the spatial configuration of these states in some grid mosaic. This raises the possibility that transition probabilities could depend on the state of neighboring patches. Grid-based models depicting connected networks of patches are also termed cellular automata, because concepts are analogous with those governing the transmission of signals through neural networks.

This chapter will introduce first the classical application of Markov models to successional changes in vegetation following some disturbance. A possible management application will be demonstrated next, which entails assigning a stocking density of livestock to maintain the vegetation in a productive state, despite environmental perturbations. Last, the construction of a spatially explicit grid model will be illustrated.

13.2   Vegetation successional dynamics

13.2.1   Forest succession

Markov matrices have been widely used to represent successional changes in vegetation following some disturbance. For example, after clear felling of a forest, the bared ground initially gets covered by grasses and herbs, and later by seedlings of pioneer trees or shrubs. At a later stage, seeds of trees that are somewhat less light demanding arrive and grow taller than the pioneers. Eventually, seedlings of shade-tolerant climax trees appear in the patch and grow slowly, but inexorably, to emerge as the forest canopy dominants. Hence four patch states may be distinguished in terms of the kinds of plants that are dominant: (i) the herbaceous stage, (ii) woodland composed of pioneer trees and shrubs, (iii) mixed woodland, and (iv) climax forest. Accordingly, the Markov transition matrix takes the form of a 4×4 square array, with the columns representing the initial states and the rows the following states (Table 13.1). Cell entries are the transition probabilities from each starting stage toward each possible next state. Note that the probabilities in each column must add up to one. A self-perpetuating state would be indicated by a value of 1.0 in the cell for self-replacement along the diagonal. The question might be how long it takes for a landscape to recover to this climax state following some disturbance, such as cyclone, flood, or clearing by humans.

Appendix 13.1 guides you into setting up a spreadsheet model of forest succession. Start with the entire area in the herbaceous state, and observe how the proportion of the landscape covered by woodland increases progressively toward the climax forest state. However, as long as a possibility of reversion from this state following a further disturbance is allowed,

Table 13.1 The structure of a Markov state transition matrix for vegetation succession from a herbaceous stage toward climax forest

Next state	Starting state			
	Herbs	Pioneer	Mixed	Climax
Herbs	$p_{h,h}$	$p_{p,h}$	$p_{m,h}$	$p_{c,h}$
Pioneer	$p_{h,p}$	$p_{p,p}$	$p_{m,p}$	$p_{c,p}$
Mixed	$p_{h,m}$	$p_{p,m}$	$p_{m,m}$	$p_{c,m}$
Climax	$p_{h,c}$	$p_{p,c}$	$p_{m,c}$	$p_{c,c}$

Note: Cell entries represent the patch transition probabilities, for example, $p_{i,j}$ represents the probability that a patch initially in state *i* will exist in state *j* one time step later.

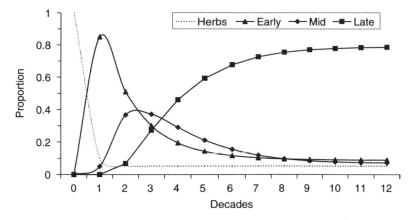

Figure 13.1 Output of model representing the changing proportion of different vegetation types during succession from a meadow toward climax forest through early-, mid-, and late-appearing tree types.

some portion of the landscape exists in earlier stages (Fig. 13.1). In reality a shifting patch mosaic would be observed as different sections of climax forest sooner or later get blown down by severe windstorms, with the overall proportion of the landscape in the different stages remaining fairly constant. These proportions are governed by the transition rates in the matrix, as you can confirm by changing the default values. Note that each time you change the value in one cell, you must adjust other entries in the same column to ensure that the transition probabilities always sum to 1.0. The program MARKOV presents the equivalent model in a different format, using the matrix multiplication function in True BASIC.

13.2.2 River channel dynamics

A state transition model was used to represent changes that had taken place in the nature of the Sabie River channel in South Africa's KNP (Rountree, Rogers, and Heritage 2000; Rogers and O'Keefe 2003). Aerial photographs showed how the riparian zone had become progressively more forested over a 50-year period. In the 1940s, the river had a predominantly wide, sandy, and rocky bed with small sections covered by reeds and a narrow fringe of woodland. Between 1965 and 1985 the area of reed-beds expanded, with reed-beds colonized in turn by trees, creating large patches of closed-canopy forest. The prevalence of forest persisted through to the late 1990s, despite a major drought followed by a moderate flood in 1996. The observed transformation of the character of the river was interpreted as possibly an outcome of the extraction of water upstream by plantation forestry and dams in the catchment, reducing water flow and flooding. Then in 2000 a huge flood occurred, removing large sections of forest and reeds and leaving a mosaic of rock, sand, and water. This brought about the recognition that the earlier open state of the river channel was largely a consequence of a similar great flood in 1925. Instead of attaining any steady state, the composition of the river channel reflected the period that had elapsed between such once in a century events.

The transition rates between the different states were assessed by dividing the river channel into a grid of 30×30 m patches in the aerial photographs. Each patch was assigned a state, and then examined in a later photograph to determine whether it had remained the same, or changed to a different state. The proportion of each patch type that had changed to some other state was thus determined over the interval between successive pairs of photographs. The actual transition rates were quite complex, changing over time and between different regions of the river channel (Rountree et al. 2000).

Appendix 13.1b guides you into setting up a spreadsheet model representing a simplified set of four river states. An approximation of the mean transition probabilities, over a decadal time step, is presented in Table 13.2a. In the absence of a flood, the general trend is toward a predominance of forest in place of open areas. By adjusting specific transition rates, you should

Table 13.2   Markov transition matrices for changes between patch states in a river channel

Next state	Starting state			
	Water	Sand	Reeds	Woodland
a. *Average transition rates in normal decades*				
Water	0.8	0.1	0.05	0.03
Sand	0.2	0.5	0.05	0.02
Reeds	0	0.4	0.7	0
Woodland	0	0	0.2	0.95
b. *Transition rates applicable in decades without a major flood event*				
Water	0.75	0.05	0.02	0.02
Sand	0.25	0.5	0.03	0.03
Reeds	0	0.45	0.7	0
Woodland	0	0	0.25	0.95
c. *Transition rates applicable in decades when major floods occur*				
Water	1.0	0.5	0.35	0.25
Sand	0	0.5	0.1	0.25
Reeds	0	0	0.55	0
Woodland	0	0	0	0.5

find that the riparian environment is highly sensitive to the likelihood that sites occupied by woodland will revert to water or sand, depending on the occurrence of scouring floods.

This deterministic model represents the usual situation in 99 out of every 100 years, when only minor floods remove a few stands of trees and reeds. When a major flood occurs, the transition probabilities are altered drastically. To represent this, you need to specify a contingent set of transition probabilities, one applicable for normal decades and the other following a major flood event (see Table 13.2b and c, and model guide-lines in Appendix 13.1c). Although on average big floods occur once in a century, because these are random events two such floods can some-times take place in close succession. At other times a prolonged period

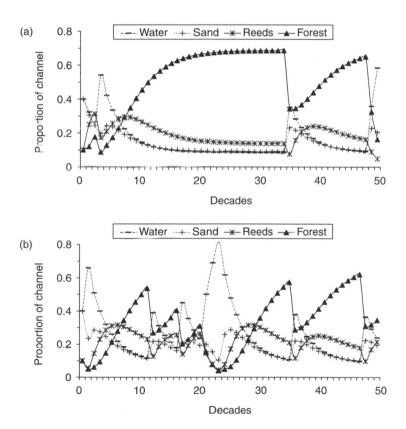

Figure 13.2 Output of model showing the impact of major floods occurring randomly once every 10 decades on the composition of a river channel, showing two contrasting outcomes arising by chance in different trials. (a) Extended interval between major floods and (b) major floods occurring in quick succession.

may pass without such a disturbance (Fig. 13.2). Hence the river state could appear very different in different centuries, without any change in the environment. Progression toward a prevalently forested state is projected to occur quite rarely. If extreme floods happen more frequently in the future, as forecast by global climate change models, the riparian zone will tend to remain relatively open. The average distribution of states is not very meaningful.

The True BASIC program RIVERSTATES represents the stochastic impact of major floods on the patch state distribution within a river channel. Compare the output from this matrix model with that produced by the spreadsheet equivalent for the same transition probabilities.

13.3   Managing savanna vegetation for livestock

Savanna vegetation dynamics has been conceptualized as "event-driven" by disturbances such as fires, interacting with weather conditions, grazing pressure, and other factors (Westoby, Walker, and Noy-Meir 1989). Since changes in state depend on the combined effect of stocking level, preceding vegetation state, rainfall and fire, management actions aimed at reversing adverse vegetation change may be effective only during windows of opportunity. Since rainfall cannot be managed, the main decision concerns what stocking level of grazing livestock (cattle or sheep) to apply.

The patch states for a savanna region can be differentiated simply as (i) open parkland, with widely spaced trees and relatively tall grass; (ii) more or less open woodland, with predominantly short grass lawns; and (iii) woody thicket, with little grass. For cattle production, the parkland state is most desirable, whereas the thicket state brings a considerable reduction in the grass forage produced. High stocking levels with cattle, and the absence of browsers, have led to extensive areas of savanna vegetation being transformed into thicket in South Africa, a problem termed "bush encroachment"(Walker et al. 1981). The mechanism is largely a reduction in fire frequency and intensity as a consequence of grazing pressure on the grass layer, which causes tall grasses to become largely replaced by shorter grasses. Grass growth, and hence fuel for fire, is also reduced during periods of low rainfall. Once the thicket state develops, it tends to persist because the sparse grass cover no longer supports hot fires.

If rainfall is high and stocking density low, the tall grass parkland can be maintained by frequent hot fires. Fires can penetrate into the margins of thickets, converting them back into parkland. Short grasses produce high quality forage for cattle, but are susceptible to invasion by thicket.

Table 13.3 Hypothetical transition probabilities between savanna patch states under different rainfall conditions, in the absence of grazing herbivores

Next state	Starting state		
	Tall grass parkland	Short grass lawns	Woody thicket
a. *High rainfall*			
Tall grass parkland	1.0	0.5	0.5
Short grass lawns	0	0.4	0
Woody thicket	0	0.1	0.5
b. *Average rainfall*			
Tall grass parkland	0.9	0.4	0.3
Short grass lawns	0.05	0.45	0.1
Woody thicket	0.05	0.15	0.6
c. *Low rainfall*			
Tall grass parkland	0.7	0.1	0.
Short grass lawns	0.15	0.7	0.1
Woody thicket	0.15	0.2	0.9

Appendix 13.3 guides you into setting up a spreadsheet representation of a savanna management model aimed at sustaining a cattle ranching enterprise (see Foran and Stafford Smith 1991 for a more sophisticated model). Suggested transition rates between patch states, for average rainfall and zero stocking, are given in Table 13.3. Obviously these rates are somewhat high for an annual time step. The starting model projects the proportions of parkland, lawn, and thicket expected from these transition probabilities. The model output becomes more interesting when allowance is made for the effect of stochastically varying rainfall conditions on these transition rates, despite the simplicity of the model (see Table 13.3). Observe how the proportion of the landscape represented by different patch states fluctuates, with parkland increasing during periods of high rainfall, and lawn plus thicket expanding during low rainfall periods (Fig. 13.3). Then introduce grazing, which promotes lawn at the expense of tall grass parkland. For simplicity, the stocking level can be categorized as low, medium, or high, with transition probabilities between the patch states modified accordingly. Compare the effects of different stocking densities on the state of the savanna vegetation (Fig. 13.4). Since the outcome depends on the rainfall sequence, repeated trials are necessary. The stocking density could be converted into some index of profitability. The potentially higher economic returns

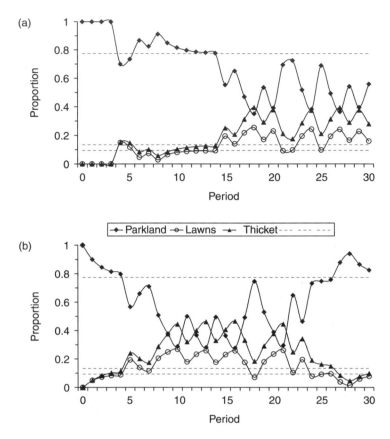

Figure 13.3 Fluctuations in the proportions of different savanna states generated by rainfall variability alone, compared with the fixed proportions that would be reached if the rainfall remained constant at the average level. Two trial outcomes showing contrasting patterns are illustrated. (a) Prevalently high rainfall initially and (b) prevalently low rainfall.

from heavy stocking are counterbalanced by the loss in forage production resulting from a reduced extent of parkland and greater prevalence of thicket. The model could be taken a step further to become a management game, with the stocking density adaptively adjusted each year in response to rainfall in the previous year, or the state of the vegetation that had resulted from prior rainfall and stocking conditions. Establish the relative productive potential of the following scenarios: (i) a ranch covered initially entirely with ideal parkland, and (ii) one acquired in a severely thicket-encroached state. Based on what you learn, establish a set of rules to govern how stocking density should be adjusted, taking into account the rainfall and the current state of the vegetation.

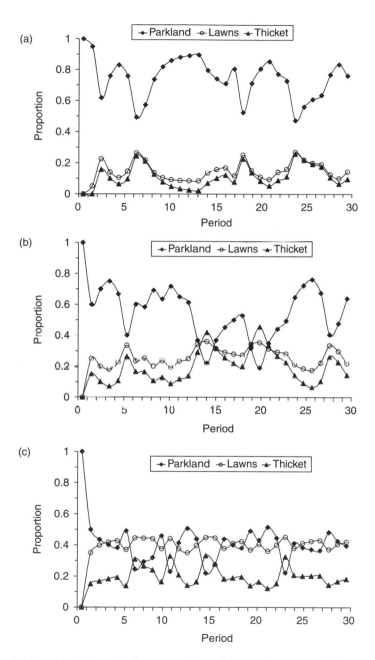

Figure 13.4 Fluctuations in the proportions of different savanna states generated by a model with variable rainfall, comparing the effects of different stocking densities of cattle. Outcomes shown for representative trials with each stocking level. (a) Low stocking level, (b) medium stocking level, and (c) high stocking level.

Textbooks on range management suggest how to establish the optimal stocking density maximizing livestock production per hectare, with neat graphs. However, the simple model you have developed reveals how difficult it is to apply these concepts in a variable environment where the optimal stocking density changes over time. Stocking levels that seem best in the short term can lead to persistent environmental changes that are ultimately detrimental for livestock production (cf. Foran and Stafford Smith 1991). Three alternative stocking strategies for coping with this uncertainty have been recognized, which may be caricatured as follows:

1 *Exploitationist* (or short sighted), that is, stock heavily to take advantage of the good times, and sell off the property to some naive city-dweller when it eventually becomes so bush-encroached as to be economically unviable for cattle ranching;
2 *Conservationist* (or risk averse), that is, keep stocking levels low to ensure that vegetation condition is maintained close to the ideal state despite periodic droughts, at the cost of lost production during good times;
3 *Wheeler dealer* (or risk prone), that is, stock heavily when rainfall is good, and reduce stock levels when drought conditions take hold to alleviate range deterioration. Since rainfall cannot be predicted, destocking only takes place after some range deterioration has occurred, so that some cost in future production is incurred. The question is, can the high profits earned during the good times compensate for the loss in production as a consequence of range degradation?

The True BASIC program SAVPATCH incorporates stochastic variation in rainfall in a similar way to the spreadsheet model outlined in Appendix 13.2. Explore the relative advantages and disadvantages of using the program version compared with the spreadsheet.

13.4 Spatially explicit grid model

The previous models considered merely the proportion of the landscape occupied by different habitat states. In practice these habitats tend to occur in blocks, and transition probabilities depend on the configuration and neighboring states of these blocks. For example, woodland patches surrounded by tall grassland are more vulnerable to being burned back than patches in the middle of grazing lawns, or extensive thickets with less edge where fires can penetrate. The reduction in edge effects is one mechanism contributing to the persistence of woodland patches once established (Jeltsch et al. 1997a). Neighborhood effects as well as consequences of patch size also arise in forest dynamics. Pioneer tree species are more successful at occupying large gaps arising from wind-throws than small gaps created merely by the demise of a single canopy tree (Alvarez-Buylla 1994). Neighborhood influences as well as those arising from patch configuration can be represented using grid-based models. In the simplest rectangular grid, each cell has four immediate neighbors.

Setting up a grid-based model in a spreadsheet is quite challenging because of the contingencies involved. It is easier to turn to the advantages of a programming language. The CD contains two alternative models, differing merely in spatial complexity. SAVGRID10 sets up a patch mosaic in the form of a 10×10 grid presenting 100 patches. It represents a hypothetical savanna landscape consisting of a mosaic of three habitat types: tall grass parkland, open woodland with short grass lawns, and woody thicket with little grass. The transition probabilities between these habitat types are adjusted by a factor dependent on the states of the four adjoining patches. Basically, patches are more likely to persist when surrounded by patches in a similar state. If the neighborhood effect is sufficiently strong, an initially random distribution of patches becomes transformed over time into larger clusters of similar habitat. Hence a vegetation pattern can develop even in a uniform environment (Jeltsch et al. 1997a). Since patch transitions are stochastic, the pattern will be different each time the model is run.

A problem in grid models is how to deal with patches located along the edges, which lack neighbors on one side. In the SAVGRID10 model, neighborhood influences are ignored for these cells. This is not unrealistic, if a defined piece of land like a ranch property is being considered. Edge effects could be eliminated in the model by joining the margins of the grid in a "double torus." To do this, assume that the sites at the bottom edge of the grid adjoin those at the top edge, and similarly for sites on the right- and left-hand edges. Wider neighborhood effects may also be considered, taking into account not just the four abutting patches, but also the additional four patches touching each square in the corners. The latter is done in the program SAVGRID20, which sets up a somewhat larger 20×20 grid encompassing 400 cells. However, the bigger the grid, the slower the program runs, especially in setting up the pictorial representation of the grid mosaic.

A grid-based model could be elaborated further to represent the concentration of particular habitats in certain regions of the landscape. For example, in moist savanna woodlands tend to develop in low-lying regions where soil water becomes concentrated, while uplands tend to support grasslands because they dry out more and thus carry hotter fires. The SAVANNA model developed by Coughenour (1992) accommodates the consequences of water redistribution within the landscape for vegetation growth, fire, and herbivore movements in different seasons. This elaborate model can be made geographically explicit by linking it with Geographic Information System (GIS) software, thereby representing the features of a specific protected area or ranch. Its seemingly realistic representation makes it a powerful aid in communicating the outcome of basic ecological processes to managers.

13.5 Overview

This chapter has been a brief introduction to a suite of modeling approaches for capturing vegetation dynamics, fire influences, and herbivore movements within a regional landscape mosaic. You should have learned the following:

1 Changing vegetation or habitat states can be modeled using Markov transition matrices.
2 Such models can represent the effects of disturbances at long intervals on the vegetation with a region.
3 Models incorporating quite simple rules can be used to explore the consequences of alternative management actions.
4 Chance combinations of events can have a persistent impact on the future state of a habitat patch mosaic.
5 Spatially explicit grid models can incorporate neighborhood influences on patch transitions.
6 Neighborhood effects can result in the coalescence of similar patches into larger blocks.
7 Grid models can be elaborated to incorporate the geographically explicit features of particular landscapes and spatially related processes taking place within them.

Recommended supporting reading

Usher (1981) describes the application of Markov matrices to vegetation succession, while Moore (1990) outlines applications of this approach in management. Korotkov, Logofet, and Loreau (2001) present a specific example applied to forest succession in Russia. Parsons et al. (2005) report how the extreme flood event of 2000 changed the state of the Sabie River channel in the KNP. Starfield (1990) outlines how very simple rule-based models can aid management decisions. Starfield et al. (1993) describe how the ecological concept of habitat states can be coupled with the artificial intelligence concept of a "frame" to address the interactive effects of elephants and fire on savanna woodland dynamics in Zimbabwe. Wiegand, Milton, and Wissel (1995) present a grid-based, "dynamic automata" model using "IF . . . THEN" rules to represent the dynamics of arid shrubland in the Karoo region of South Africa. Jeltsch et al. (1997b) used a grid-based model to depict the pattern established by grazing impacts concentrated around waterpoints, while Weber et al. (1998) developed a similar model to represent vegetation responses to spatial heterogeneity in grazing impacts across a wider landscape.

Programs on the accompanying CD

MARKOV; RIVERSTATES; SAVGRAZE; SAVGRID10; SAVGRID20

Exercises

1 Take the forest succession model developed in Section 13.2 and parameterize it more realistically to represent some particular example where the transition rates have been measured.
2 Transform herbivore stocking rates in the savanna model into a more realistic measure of economic returns, and use this model to estimate the economically optimal tactics for different environmental conditions.
3 Formulate a more realistic grid-based model of savanna state dynamics, drawing from some of the references given above, and note how the state distribution differs from that projected by a nonspatial model with the same transition frequencies between states.
4 Develop the concepts outlined by Starfield et al. (1993) into a model for managing elephant impacts on vegetation state, by manipulating both elephant abundance and fire frequency.

14

Habitat suitability models

Adaptive behavior

Chapter features

Topics

Habitat choice; carrying capacity; behavior-based models; evolutionary optimality; resource types; buffers

14.1 Introduction

Habitat loss or deterioration has become a major issue threatening the viability of many populations. Much of this loss has been due to human impacts, but climate change has loomed as an additional influence to be recognized. The challenge is to anticipate the consequences of such threats, so that they can be ameliorated, or in some situations deflected, particularly where rare or valuable species are involved. Since experiments are rarely possible in such circumstances, models based on existing knowledge must be invoked.

Habitat dependency is commonly assessed statistically by fitting resource selection functions to the use of alternative habitats relative to their availability (Boyce and McDonald 1999; Manly et al. 2002). The factors associated with these preferences may be identified through relating use to various factors present in these habitats. This approach can be useful in indicating differences in habitat suitability, but is of limited value in projecting the consequences of habitat loss, particularly for the population trend. A more fundamental approach is needed, identifying the specific contributions of available habitat types to population performance.

Performance can be expressed either in terms of the population size maintained, or the rate of increase shown when the population is below its potential abundance level. Habitat suitability depends most fundamentally on the food resource base, specifically the accessibility, availability, nutritional value, and seasonal variability of different food types. The population performance can be modified by the relative vulnerability to predation or parasitism in each habitat, as well as by physiological stresses incurred during extreme weather. Animals commonly shift between habitats seasonally, and populations may spread across a wider range of habitats when population density is high. The available habitats may be contiguous, or widely separated in space. Some birds migrate between breeding and overwintering habitats thousands of kilometers apart. The cost of travel becomes a factor affecting habitat occupation in such situations.

In the absence of experimentation, principles of evolutionary optimality can serve as a guide to the likely responses of organisms to changed or novel circumstances. Individuals that have made the appropriate decisions in the past have passed on their genes, which may predispose their descendants to respond appropriately to future change, provided conditions are not too different. These principles have been identified and tested in the field of behavioral ecology (Lima and Zollner 1996; Krebs and Davies 1997). The challenge is to apply them to identify the population consequences of novel circumstances. Real animals approach the idealized outcome of optimality analysis somewhat imperfectly, due to limitations in information or other factors. The models nevertheless identify the best outcome possible in the prevailing circumstances, given the assumptions incorporated into them.

The pattern of habitat use depends on how intra-specific competition is expressed (Sutherland and Norris 2002). If competition arises largely indirectly through resource depletion, animals are likely to concentrate in the currently best habitat, then shift en masse to another habitat after a critical threshold of depletion has been passed. This is the pattern shown by species that tend to aggregate in large herds, like many waterfowl and ungulates. If direct interference through aggressive interactions occurs, for example, through territorial exclusion, the more dominant individuals tend to occupy the best habitats, with younger or weaker individuals relegated to less favorable habitats. For example, great tits breed preferentially in woodlands, but with young birds reproducing for the first time displaced into hedgerows (Krebs 1971).

In this chapter, you will be introduced first to a generic habitat use model for migratory waterfowl occupying different habitats during the overwintering period. The aim of the model is to assess the relative effects of losses or restrictions in the availability of these habitats on the abundance of the population and consequent impacts in farmlands. It illustrates how adaptive habitat choices based on relative gains can be incorporated into such models in a simple way. A more elaborate model assessing the capacity of different regions, defined by their vegetation composition, to support a mammalian herbivore will then be developed. It illustrates how resource components exploited at different stages of the seasonal cycle contribute toward habitat suitability. Such models could help avoid some of the failures that have occurred when animals have been introduced into habitats that superficially have seemed suitable (Owen-Smith 2003).

14.2   Shifting habitat use by overwintering geese

Many goose species breed in arctic latitudes, and then migrate to spend the winter months further south in Europe or North America. For example, brent geese (*Branta bernicla*) nest on the Taimyr Peninsula in northern Russia, and overwinter along the coastlines of western Europe. This largely marine species depends on eelgrass (*Zostera* spp.) growing in the intertidal zone as its staple food resource, and thus concentrates in coastal habitats. However, after the population had grown following a hunting ban, the geese fed also on grasses in pastures and less profitable natural grassland or arable lands during the winter and early spring periods, leading to conflicts with farmers (Pettifor et al. 2000a). A second species, the Barnacle goose (*Branta leucopsis*), nests on Svalbard Island, Norway, and spends the winter in a localized area around the Solway estuary in Britain. It was reduced to only around 300 birds during the 1940s, but has since increased to over 23,000. These geese spend most of the winter grazing in salt-marsh and adjoining pastures, with occasional forays into nearby arable fields. During their spring return to Svalbard, they stop over briefly on islands serving as staging areas.

The model to be developed addresses the consequences of constrictions or losses among the alternative habitats available to migratory waterfowl such as these goose species. Basic concepts have been extracted from the detailed models outlined by Pettifor et al. (2000a). However, the model you will develop is a greatly simplified version. It covers only the 6-month wintering period, and competition is expressed solely through resource depletion. Habitats are assumed to be contiguous, so that birds can move freely between them with little cost. Three alternative habitats will be considered, including (i) the natural habitat, for example, salt-marsh, potentially providing the highest food gain; (ii) an alternative habitat, such as grazed pastures, used after food availability in salt-marsh has been depressed sufficiently; and (iii) a low-value habitat, such as arable fields, serving as a buffer after food availability elsewhere has been greatly reduced. No plant regrowth is incorporated, so that the pool of available food built up during the preceding summer period becomes progressively reduced as a result of consumption. Appendix 14.1 provides guidelines for setting up such a model in a spreadsheet.

The weekly or daily value of feeding in a particular habitat is assessed simply by the intake or "functional" response of the birds to the current food availability. Hence the value of a habitat declines as food becomes depleted, and more rapidly if the abundance of the geese is higher. At some stage gains from the initially best habitat become less than those from the alternative habitat, and the birds switch habitats. Hence rate maximization of food intake is the criterion used to assess habitat value. Since the habitats are contiguous, birds can switch back to the initially better habitat following depression of the resources offered by the alternative habitat. Eventually food availability in both these habitats becomes so low that it becomes worthwhile for the birds to occupy the buffer habitat, which serves to slow their rate of starvation toward the end of winter.

The basic model indicates the stages at which switching between the habitats occurs, dependent on the parameter values governing the relative intake responses in these habitats, and the abundance of the goose population (Fig. 14.1). At low abundance the geese can remain in the best habitat throughout the winter because resources suffer little depletion. Higher population levels of geese lead to earlier switches between habitats. In this way a greater population of geese leads to increased conflicts with farmers.

To assess the consequences of changes in habitat availability for the abundance of the geese, the rate of food gain obtained by the birds through the winter period needs to be transformed into a potential rate of population increase or decline. We assume that resource shortfalls during winter increase the vulnerability of the geese to mortality during this period, or at least restrict the build-up of the fat reserves needed for successful reproduction. The effects can be revealed by deriving the resultant density-dependent relationships for population growth (Fig. 14.2). Reducing the extent of the most favorable habitat has a greater effect on the population size attained at zero growth rate than corresponding reductions in alternative habitats, as might be expected in the real world. However, almost eliminating the alternative habitat greatly lowers the size of the goose population that can be supported. This suggests the contribution that agricultural fields might have made to the observed expansions in goose numbers. The contribution

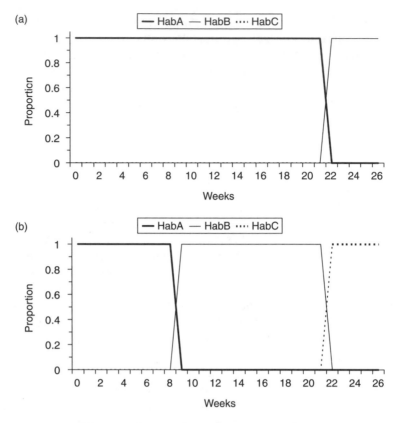

Figure 14.1   Model output showing changes in the use of three alternative habitats over the winter period, as affected by the abundance of the goose population, labeled A, B, and C in order of their suitability. (a) Low abundance level, and (b) high abundance level.

of the buffer habitat seems small under average conditions because of its low returns, but its importance is revealed when the effects of annual variation in food production within the preferred habitat are considered (Fig. 14.3). The population established at its zero growth density under average conditions declines more in poor years when the buffer habitat is removed than when it is present.

This generic model could be elaborated to incorporate realistic information about particular goose populations, drawing on the detailed data presented by Pettifor et al. (2000a) for brent geese and barnacle geese in Europe. The competitive influence of other waterfowl species overlapping in resource use could also be examined, following the example of Sutherland and Allport (1994). Models incorporating competitive interference among birds differing in dominance status have been formulated in great detail for

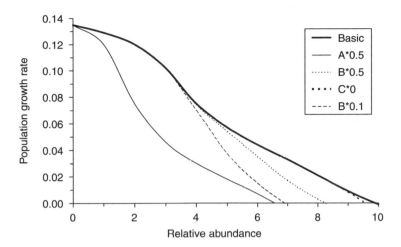

Figure 14.2   Modeled population growth rate in relation to population density as affected by habitat availability for a hypothetical goose population. Habitat A potentially yields highest gains, habitat B intermediate gains, and habitat C lowest gains. In the basic situation (broad line), all three habitats are equally available. Other lines show the effect of reductions in the relative availability of particular habitats.

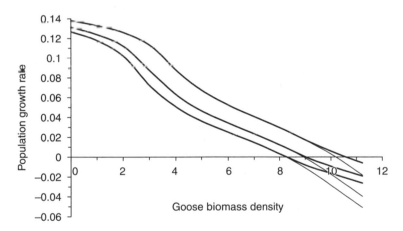

Figure 14.3   Modeled consequences of annual differences in food production in the best habitat (range 80–110 units) on the density-dependent growth rate of a hypothetical goose population, as influenced by the presence (broad lines) or absence (thin lines) of the buffer habitat C.

oyster catchers foraging on mussels and other shellfish in tidal flats along the British coastline (see Goss-Custard et al. 1995a,b).

A counterpart habitat choice model written as a program in True BASIC is on the CD, representing some generic herbivore. This model compares the outcome of two alternative choice algorithms. One is based on the idealized optimality principle with consumers choosing the habitat that maximizes nutritional gains at each time step, as adopted in the spreadsheet model for the goose population. The other considers an alternative model in which the population is distributed among the available habitats in proportion to the relative gains that each offers (called "matching"). The latter is believed by some observers to better represent what is observed in real populations.

14.3   Habitat suitability for a browsing antelope from vegetation composition

This section draws on a model developed to explain the widely contrasting abundance shown by a large browsing antelope, the kudu, in different regions of savanna vegetation in southern Africa (Owen-Smith 2002b, chapter 11). Typical savanna bushveld supports kudu densities of 2–3 animals km^{-2}, but some savanna regions have no resident kudu herds, while in succulent thicket vegetation in the eastern Cape kudu density reaches over 10 animals km^{-2}. The model assesses how features of the vegetation composition affect the density of the herbivore population that can be supported, taking into account the food limitations that arise during the dry season.

The basis of the model is the food intake function and the conversion of the food ingested into nutritional gains relative to metabolic requirements. This nutritional balance determines the susceptibility of the animals to mortality, whether directly through starvation or as a consequence of the greater movements and higher risks that they incur in the process of seeking sparse food. The vegetation features considered relevant include the abundance, nutritional yield as digestible dry matter, consumption rate, and seasonal foliage retention characterizing a set of plant types. Changes in the herbivore population are considered merely in terms of the aggregate biomass dynamics, without taking into account any subdivision into age or sex classes.

A key aspect of the model is to allow for changes in food selection by the herbivores in response to the reduction in availability of the preferred food types through both consumption and intrinsic attrition via leaf fall. Dry season conditions are considered, when plant regrowth ceases, since this is the limiting period of the year nutritionally. If the animals continued searching for just the preferred food types under such conditions, they would starve from lack of food. If they ate all food types indiscriminately, they would starve through not being able to digest much of the fibrous material. Somewhere between these extremes lies the optimal diet, that is, the combination of food types maximizing the nutritional gain and thereby minimizing susceptibility to mortality under conditions where nutritional deficits develop.

Hence, rather than considering selection between alternative habitats as in the case of the goose model, the model is focused on selection among the food types intermingled within the same habitat. Obviously, the best food type should always be eaten when encountered. The question is whether to ingest the second-best food type as well, plus the third best, and so on down the rank order of value. The model calculates the consequences of widening the dietary range to include additional food types, and identifies the combination of food types giving the highest daily nutritional gain, in digestible dry mass. This requires some basis for ranking the food types in the order of their effective nutritional value, which is assumed to be indicated by their digestible dry matter content. The mortality rate is assumed to be inversely related to the nutritional gain relative to requirements. Calculations are done on a weekly basis through the dry season period, and the averaged susceptibility to mortality assessed and counterbalanced against the recruitment potential from offspring produced during the summer or wet season. Heightened competition for food arising from a greater population density is expressed simply through a faster depletion of the favored food types, and hence an increase in the potential mortality loss. These calculations can be used to plot the potential population growth rate as a function of increasing population density. The density level at which the net population growth rate becomes zero is, by definition, the carrying capacity or population density that can be supported by the seasonally changing food resources. The higher the carrying capacity, the more suitable the habitat is for the species.

Estimates for input into the model were obtained from a detailed study conducted on habituated kudus that allowed close-range observations despite ranging fairly freely within a 210 ha enclosure (Owen-Smith and Cooper 1987, 1989; Owen-Smith 1993a, 1994). The data collected included the accessible leaf biomass produced, contents of various nutrients in the leaves, bite sizes and biting rates obtained while feeding, and foliage loss through both leaf fall and consumption through the dry season. These derived measures were distinguished among plant types, represented by sets of species with similar characteristics, which differed in their representation in different savanna types (Table 14.1).

During the dry season months plant growth ceases, and the biomass of the food available to kudus declines drastically through leaf fall and senescence of forbs (herbaceous plants besides grasses), together with the effects of consumption (Fig. 14.4). The model calculates the consequent

Table 14.1 Features of the plant types available to kudus in three savanna regions

	Nutritional conversion c	Initial biomass F (g/m^2)	Eating rate e (g/min)	Leaf attrition rate (/week)	Onset of leaf fall (week)
a. *Broadleaf savanna*					
Forbs	0.72	3	5	0.15	4
Palatable deciduous	0.65	8	5	0.15	10
Palatable spinescent	0.70	1.5	2.5	0.15	8
Palatable evergreen	0.58	1.5	5	0	—
Unpalatable deciduous 1	0.50	1.5	5	0.15	14
Unpalatable deciduous 2	0.35	19	5	0.15	14
b. *Acacia savanna*					
Forbs	0.75	6	5	0.15	4
Palatable deciduous	0.65	0.5	5	0.15	10
Palatable spinescent A	0.70	4	2.5	0.15	8
Palatable spinescent B	0.72	12	1.5	0.15	8
Unpalatable deciduous	0.50	3	5	0.15	14
c. *Succulent thicket*					
Evergreen A	0.65	8	5	0	—
Spinescent evergreen	0.7	8	2.5	0	—
Evergreen B	0.60	8	5	0	—
Evergreen C	0.5	8	5	0	—
Evergreen D	0.40	8	5	0	—

reduction in nutritional gains by the kudus relative to requirements. The susceptibility to mortality accelerates sharply once submaintenance intake levels are reached.

Using vegetation data for the broad-leaf savanna where the study was conducted, the model projected zero population growth at a kudu density of around 2–3 animals km^{-2} (averaging 140 kg each in body mass), closely matching the population levels typically observed in such vegetation (Fig. 14.5). In contrast, there were no resident kudu herds in regions

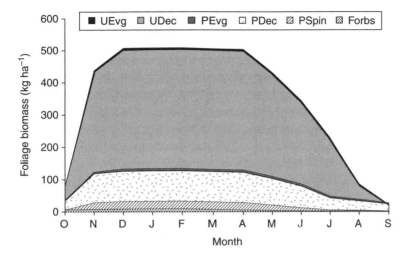

Figure 14.4 Seasonal changes in foliage biomass available to kudus, partitioned among plant types for broad-leaf and fine-leaf savanna combined (data from Owen-Smith and Cooper 1987; Owen-Smith 1994).

where umbrella thorn acacias predominated. The features distinguishing umbrella thorn savanna are the restricting eating rates offered by the tiny leaves surrounded by thorns of the predominant acacias, and the lack of evergreen trees and shrubs retaining leaves through the dry season (Table 14.1b). In these circumstances, the model projected a carrying capacity for kudus not much above zero (Fig. 14.5), despite a total production of food during the wet season similar to that for broadleaf savanna. Vegetation measurements were lacking for the succulent thicket prevalent in the eastern Cape. Hence I assumed that the forage biomass produced within the height reach of a kudu differed little from that in savanna, but with the vegetation made up entirely of evergreen species retaining leaves through the dry season (Table 14.1c). This change elevated the kudu density supported toward the observed densities in this habitat, which exceed 10 animals km^{-2} (Fig. 14.5). Overall, the model emphasized the importance of vegetation components retaining forage through the dry season for determining the browser population that can be supported, because of the consequences of nutritional shortfalls for mortality during this critical period.

This model can be used to explore the consequences of specific adjustments to vegetation composition, for example, increasing or decreasing the evergreen component in broadleaf savanna vegetation. A surprising finding was that eliminating unpalatable tree species, which contain high tannin contents restricting their digestibility, substantially reduced

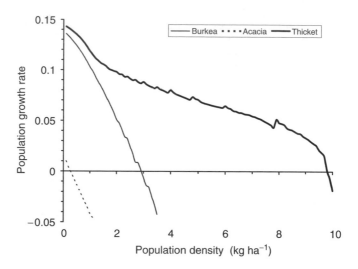

Figure 14.5   Model output projecting the population growth rate of kudus at changing density levels, comparing three habitats: broad-leaf savanna, acacia savanna, and evergreen thicket (from Owen-Smith 2002b).

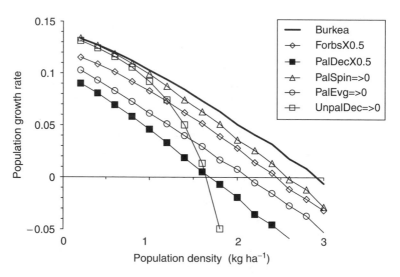

Figure 14.6   Output of the browser model showing the effects of adjusting the abundance in the vegetation of particular plant types (from Owen-Smith 2002b).

the kudu density supported (Fig. 14.6). These species flushed new foliage before the end of the dry season, shortening the period over which the herbivores persisted on a starvation diet, while also forming a large proportion of the forage biomass. Hence they served as a key bridging resource during this transitional period of the year.

Besides the simple HABCHOICE model, you will find two more elaborate versions on the CD:

1 HABGAINS calculates the weekly food intake and nutritional gains of a large herbivore through the dry season and consequent susceptibility to mortality;
2 HABPOPUL projects how the potential population growth rate varies with stocking density, used to produce the plots shown in Fig. 14.5.

14.4   General principles

The models developed in this chapter illustrate how the capacity of the vegetation to support a herbivore population depends not only on how much food is produced during the growing season, but also upon the retention of edible vegetation components through the dormant season when vegetation growth ceases. A counterpart model showing how components of the grass forage resource contribute to supporting the stocking density of grazing livestock was introduced in Chapter 12. The important principle incorporated in these models is that the herbivores adjust their selection of habitats and of food types offered within them in response to changing food availability. Hence principles of adaptive behavior need to be incorporated into models to represent these shifting choices.

Particular vegetation components differ in the contributions that they make toward sustaining herbivores at different stages of the seasonal cycle. *Staple* food species provide the major portion of the diet while they retain foliage. These were constituted by palatable deciduous trees in the case of kudus. *High quality* food types contribute importantly to nutritional gains and hence reproductive success during the growing season. These were represented by forbs and to some extent the spinescent acacias for kudus. *Reserve* food types become important after the availability of the staple foods declines, as a result of consumption or inherent attrition. Evergreen or semi-evergreen trees and shrubs constituted this category for kudus. Last, *buffer* or *bridging* food types gain importance during periods when little other food is available. Unpalatable species flushing new leaves early served this function for kudus. Similarly, the models for geese outlined by Pettifor et al. (2000a) recognize the contribution that particular habitats make toward supporting the populations at different stages of the year, specifically in breeding, wintering, and staging areas.

A fundamental uncertainty in rate-maximizing models is how food gains from earlier times are carried over as stored reserves to offset later shortfalls. Pettifor et al. (2000a) sidestepped this issue in their physiologically explicit model by assessing mortality in relation to some minimum body mass for survival. A similar principle was used by Illius and Gordon (1998) to model the mortality incurred by Soay sheep over the annual cycle. This approach neglects the greater exposure to mortality that animals incur when they forage for longer to avoid starvation, for example from the risk of predation. While the models developed in this chapter relate habitat suitability only to food resources, relative security against predation is an important additional consideration affecting habitat choice.

The rate-maximizing principle governing resource selection in the above models has also been challenged as an unrealistic ideal. Real animals make imperfect choices based on what they experience rather than the perfect assessment of relative gains incorporated into the models. Nevertheless, the models identify the upper limit to performance set by the environmental conditions included in the model. Alternative choice algorithms could be used, as illustrated in the HABCHOICE program. Beside rate maximizing or "matching," competitive interactions having immediate effects on resource gains dependent on population levels within each habitat could be represented (Rosenzweig and Abramsky 1997). This requires merely altering the food intake function to incorporate density-dependent interference (see Section 5.4).

14.5   Overview

From this chapter, you should have learned the following:

1   The abundance level of a population is set fundamentally by the available habitat types and food resources on offer within these habitats.
2   Use of these habitats and food types changes at different stages of the seasonal cycle.
3   Projecting the patterns of habitat and resource use under different conditions requires the incorporation of decision rules, which invoke proxy measures of evolutionary optimality.
4   Minor resources and habitats can make major contributions toward supporting populations through critical periods.
5   Models that ignore the adaptive response of consumers to changing conditions are likely to give misleading projections of the abundance levels sustained in different circumstances.

Recommended supporting reading

Norris (2004) presents a strong case for the use of behavior-based models incorporating fundamental evolutionary principles in threatened species conservation (see also Pettifor, Norris, and Rowcliffe 2000b; Norris and Stillman 2002; Sutherland and Norris 2002). The important point emphasized is that information from the past is an unreliable guide to the likely dynamics of populations in the novel environments of the future. Sutherland (1996) provides examples showing how behavioral concepts can be applied in population ecology, largely for birds. A specific application for shorebirds addressing the consequences of shellfish management is outlined by Stillman et al. (2001). Owen-Smith (2002b, 2005) advocates a metaphysiological modeling framework linking adaptive behavior to the consequences for population dynamics, with large mammalian herbivores as the focus. Persson and De Roos (2003) describe a physiologically structured model for fish, which incorporates predation risk as an additional factor in habitat selection.

Programs on the accompanying CD

HABCHOICE; HABGAINS; HABPOPUL

Exercises

1 Expand the goose habitat use model to incorporate additionally the timing of departure from the breeding range and the use of staging habitats en route, extracting the needed information from Pettifor et al. (2000a).
2 Parameterize the generic habitat use model for geese that you developed for some specific waterfowl population, using data from the literature (e.g., Durant et al. 2003).
3 Experiment with the habitat suitability model for browsing kudus by changing the relative availability, quality, or other features of the resource types present, and establish how much difference this makes to the population that can be supported.

15

Reconciling models with data

Statistical diagnosis

Chapter features

Topics

Information theory; alternative hypotheses; model selection; declining populations; Akaike Information Criterion; predictors; relative likelihood; coefficient of determination

15.1 Introduction

In the preceding chapters you have been introduced to a variety of modeling approaches. You have learned that modeling can be helpful even when no data are available, through helping conceptualize a problem and by identifying the kind of information that would be most useful. Having obtained such data, there may still be uncertainty about which factors were most influential, and how they operate. Recognizing that stochastic processes are pervasive, results obtained could be partly a chance outcome of such processes. To exclude this possibility as being highly unlikely, you will need to draw upon procedures that have been developed in the field of statistics.

The statistical approach traditionally followed has been to assess the probability that the data could have been obtained under some null hypothesis. If the statistical test applied indicates that the probability of obtaining these data, or data even more extreme, is less than some preassigned level (generally 1 in 20) had the null hypothesis been valid, there is convincing evidence that the observed pattern was not due to chance. This is called the "frequentist" approach, because it is based on assessing the frequency, or proportion of repeated trials, that could have produced such extreme results given some assumed probability distribution around the mean.

Nevertheless, having rejected the null hypothesis, what alternative explanation should we believe, given that many possible factors could be influential? Rather than jumping to the conclusion that our favored explanation is valid, we should consider possible alternatives. These may be formalized as alternative models, incorporating specific hypotheses about the factors postulated to be influential. The challenge is to reduce the candidate set of possible factors to some more limited set of the most important influences. Hilborn and Mangel (1997) characterized this approach of multiple hypothesis testing as entailing *ecological detective work*.

An objective basis for selecting the best model from among a set of candidates is provided by formal information theory. Complete information would enable us to specify exactly the outcome to be expected, but full truth is an unattainable ideal for complex biological systems. Given the limited data that are available, how many contributory influences can be supported? Including more factors in the model might seem to make the model fit the data better, as judged by an increase in the "coefficient of determination"(R^2) if fitting is by least squares. However, this is at the cost of less precise estimation of the influence of each factor, as indicated by wider confidence limits for their effects. How do we judge which model is best supported by the data, without including more factors than are warranted?

This chapter will illustrate how the Akaike Information Criterion (AIC), derived by a Japanese statistician, can be used to identify the best approximating (or most parsimonious) model, while recognizing the relative support for alternative models. This index is based on the Kullback–Leibler distance, which measures the amount of information that is lost when a given model is used to approximate the full truth. The problem to which it will be applied concerns a diagnosis of the factors primarily responsible for the declining population trends shown by certain less-common antelope species in a large protected area, threatening the local extirpation of some of these species. Caughley (1994) highlighted the lack of any coherent theory to explain population declines, and thereby enable remedial action to be taken before population viability becomes threatened. Experimentation is rarely possible in such situations, hence the most likely cause or set of causes must be inferred given the data that are available. To do so requires a combination of theory, model construction, and statistical assessment, as demonstrated below for the selected case study.

15.2   Model selection statistics

An explanatory model has three basic components: (i) the factors hypothesized to be influential, that is, the set of predictors; (ii) how these factors exert their effect, that is, the connections and form of the functional relations; and (iii) potential interactions between factors, that is, how the effect of one factor depends on another. In population dynamics, the response variable is generally expressed as the change in abundance from one time step to the next. Various factors could cause the population to increase or decrease – alterations in food resources, predation, disease, competition, weather, and the like. The model is generally expressed as a regression, relating changes in abundance to quantified variation in one or more candidate predictors. Alternatively, spatial differences in mean abundance could be related to various habitat factors, leading to a classical analysis of variance (ANOVA). If distribution patterns are expressed in terms of the frequency of occurrence in different habitat types, an elaboration of contingency tables called log-linear analysis (because the data are log-transformed to make the relationships linear) would be appropriate. These are variants of General Linear Models (or Generalized Linear Models, if a nonnormal error distribution is invoked), differing simply in whether the predictor and response variables are measured continuously or categorically.

Whatever the data structure, the statistical output will indicate the residual variation in the data not accounted for by any particular model. With classical ANOVA or least-squares regression, this is shown by the "residual sum of squares." If maximum likelihood procedures are followed, the negative log of the likelihood associated with the most likely estimates of the model parameters, given the data, indicates the departure from full truth. In calculating the AIC, an additional penalty is added for the number of parameters estimated in fitting the model. For maximum likelihood estimation, the formula is

$$AIC = -2LL + 2K,$$

where LL is the value of the maximized log-likelihood and K is the number of predictors in the model, plus one for the regression intercept if estimated and one for the estimate of the variance. If a least-squares procedure is adopted, the equivalent formula is

$$AIC = n \log_e(RSS/n) + 2K,$$

where RSS is the residual sum of squares and n is the sample size. The model best fitting the information in the data is that yielding the smallest (or most negative) AIC. If the sample size is small relative to the number of predictors considered (i.e., $n/p < 40$), an additional correction term should be applied:

$$AIC_c = -2LL + 2K + 2K(K + 1)/(n - K - 1),$$

where AIC_c is the corrected Akaike Information Criterion.

The absolute value of AIC (or AIC_c) has little meaning; it is the relative values for the alternative models that are of interest. Hence the model with the smallest AIC is commonly assigned a value of zero, and the relative AIC difference, ΔAIC, for other models calculated by subtracting the AIC value of the best model. Models with similar AIC values are almost equally well supported by the data. The weight of support for a particular model w can be calculated from its relative contribution to the ΔAIC values:

$$w_i = \frac{\exp(-\Delta AIC_i/2)}{\Sigma \exp(-\Delta AIC/2)},$$

where the denominator is summed across all of the models considered. Hence the relative likelihood of model i versus model j is w_i/w_j. These relative likelihoods can be scaled up so that the most likely model receives a value of 1, and other values are proportional to this. A model differing by 4 units in AIC is 13.5% as likely as the best model, while one differing by 7 units is only 3% as likely.

You should recognize that the calculated AIC values would change if a different data set was obtained, and affect the estimated likelihoods of the alternative models. Hence all models receiving sufficiently substantial support (within 4–7 AIC units of the best model) should be regarded as candidates for becoming the best model for another data set. The set of models considered should be restricted to those judged as biologically plausible. Subjecting a large number of models to testing using the AIC represents mere "data dredging" rather than scientific assessment.

15.3   Diagnosing the causes of antelope population declines

15.3.1   Initial approach

Green (2002) outlined sequential steps that should be followed in diagnosing the factors responsible for declining population trends, as well as some pitfalls, with birds in mind (Box 15.1). These same points are relevant to explaining the drastic decreases in abundance shown by several antelope species in South Africa's Kruger National Park (KNP), threatening a reduction in the diversity of the large herbivore assemblage. Establishing the causal factors responsible is not merely of academic interest, but essential for identifying the remedial actions needed. While uncertainty about the causal processes retards action, populations remain low and subject to the various influences that threaten the viability of small populations. Hence the basic questions to be addressed are (i) what caused the decline in abundance of these species? and (ii) what should be done to rectify this situation?

The KNP is a large ($20,000\,\text{km}^2$) protected area, which had been entirely fenced since 1975 (sections of the fence have recently been removed). Unusually good data were available recording the population trends of all of the larger ungulate species both before and through the period of the declines, from parkwide aerial counts conducted annually between 1977 and 1996. These were supported by surveys documenting the age and sex structure of these populations from 1983 onward. The analysis in this chapter will be focused on three species: (i) roan antelope (*Hippotragus equinus*), which declined from 300 to 450 animals prior to 1986, to only 25 free-ranging individuals by 1996; (ii) sable antelope (*Hippotragus niger*), which dropped from over 2,000 to around 500 animals; and (iii) kudu, which decreased from a peak population of over 10,000 to around 3,000 (Fig. 15.1). For a more complete assessment encompassing additional ungulate species, refer to Harrington et al. (1999), Ogutu and Owen-Smith (2003), Owen-Smith et al. (2005), and Owen-Smith and Mills (2006).

The factors potentially responsible for the population declines included (i) direct effects of below-normal rainfall, (ii) habitat deterioration as a result of the low rainfall, (iii) an outbreak of some disease, (iv) displacement by a competitor, (v) elevated predation, or (vi) illegal or uncontrolled hunting. The available data included changes in aerial census totals for other large ungulates, and rainfall records collected at various ranger stations scattered throughout the KNP. No information was available on vegetation changes, nor on predator populations.

Illegal hunting ("poaching") could be eliminated from consideration because the KNP is well patrolled and declines were not localized in boundary regions where hunters might penetrate. While the KNP is subject to periodic outbreaks of anthrax, which in the past have caused mortality among roan and kudu, the occurrence of anthrax-related deaths is closely monitored, and none were recorded at the time when the population declines were initiated.

Box 15.1   Steps in diagnosing the causes of a population decline (modified from Green 2002)

Step	Questions	Pitfalls
1. Assess the weight of the evidence indicating a declining trend	Is the decline local or regional? When did the decline commence?	No data are available before the decline was recognized Decline might be part of normal pattern of fluctuation
2. Assemble available data for the species and other similar species	What are the habitat requirements? What are the food resources? What affects the production of these resources? What are the predators? Are there potential competitors?	Only superficial information is available Information from elsewhere may not be applicable to this population
3. Establish whether the decline is associated with poor reproduction or low survival	At what stage in the life history is the loss occurring?	Data on vital rates are unavailable, or unreliable Similar species may differ in the demographic profiles
4. List all plausible factors that might be responsible for the declining trend	Which factors are most likely to influence reproductive success and which survival in later stages?	The list could become very long
5. Establish what information is available for the factors that could be involved	How reliable is this information for the area of concern?	The available information is for a different place No information is available for some salient factors
6. Establish the statistical association between the population trend and candidate factors	What is the strength of the correlation with particular factors?	some factors omitted due to lack of information Effect depends on interactions between factors Spurious correlations may be misleading
7. Assemble a reasonable set of alternative models and assess how well each is supported by the data	Which factors should be included in models? What functional form does their influence take?	Too little data to support more than one or two factors Several models receive very similar support
8. Incorporate the identified influences into a simulation model to establish the magnitude of the remedial action required	How much change in a factor is needed to alter the population trend?	The model does not include all relevant interactions The model does not allow for some external influences
9. Adjust the factors identified as important and observe the effect on the population	How much change is needed to make a difference?	Another factor besides that adjusted could be responsible for the changed trend

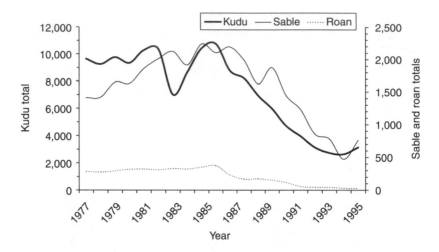

Figure 15.1   Changes in the count totals for kudu, sable, and roan antelope over the entire KNP between 1977 and 1996, from South African National Parks files. The totals have been extrapolated for early and later years when only sections of the park were covered.

15.3.2   Exploratory pattern seeking

KNP scientists had noted that the decline of the roan antelope population was associated with an increase in the abundance of zebra in the northern section of the park, where the roan mainly occurred (Fig. 15.2). This suggested that the zebra could be outcompeting the roan. However, the decrease in roan numbers seemed more extreme than could be explained by the doubling in zebra abundance. Moreover, kudu and sable had also declined, although not restricted to this region.

The roan decline was initiated a few years after the severe El Nino-related drought of 1982/3, in a year that was not quite as dry (Fig. 15.3). A further severe decline was associated with the also severe 1991/2 drought. Hence it seemed that low rainfall affecting grass production might have contributed to the negative population trend, although not immediately. Further analysis of the rainfall data indicated that rain received during the normally dry season months had been exceptionally low from 1987 onwards. This suggested that it might have been low food quality as a result of few green leaves remaining during this critical period of the year that was largely responsible for the negative population trend after 1986. Furthermore, the kudu model relating stage-specific survival rates and hence population trends to rainfall (see Chapter 6) indicated that low annual rainfall alone could not account for the declining trend shown by this species across the KNP after the conclusion of my study (Owen-Smith 2000).

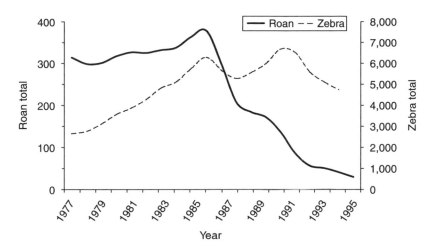

Figure 15.2   Relation between the decline in the number of roan antelope counted in the northern section of KNP and the number of zebra counted in this region.

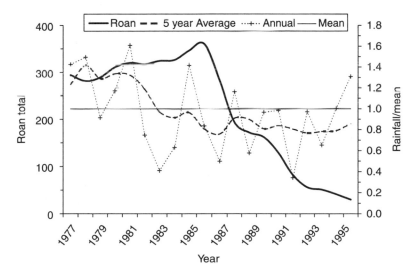

Figure 15.3   Relationship between changes in the roan antelope total and changes in rainfall in the northern section of KNP.

In interpreting the causal mechanisms, it helps greatly to know whether the decreasing population trend was due to poor reproduction or low survival of mature animals. Through reconciling the annual population change with recruitment indicated by the proportion of calves relative to mothers, from the sex and age classifications (see Owen-Smith et al. 2005 for the procedure), it became evident that the declines were largely

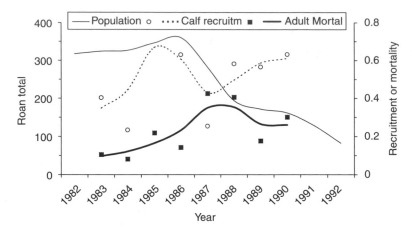

Figure 15.4   Annual changes in calf recruitment indicated by the cow : calf ratio in the population, and the projected mortality rate for the segment comprising yearlings plus adults from the accounting model for the roan antelope (symbols show annual estimates, lines the smoothed trends).

a result of elevated adult mortality (Fig. 15.4). For roan antelope in particular, calf recruitment was low during the years with low rainfall, but rebounded when rainfall improved. However, mortality among the adult segment tripled from an initial loss of about 10% per year to over 30% after 1986, and remained high.

In general adverse nutritional conditions affect the survival of young animals much more strongly than that of adults (Gaillard et al. 1998). This suggested that some additional factor affecting adult survival, besides low seasonal rainfall, was involved. Hence a possible increase in predation pressure needed to be considered, in particular, by lions which prey to a large extent on adult ungulates. However, lions cannot be counted from the air (they are mostly hidden under bushes during the day), so no direct information on changes in lion abundance were available. Nevertheless, a check of ranger's diaries for the sections where the roan decline had occurred revealed that both sections had recorded increases in lion sightings through the period of the decline. The implication was that the increase in zebras had expanded the prey base for lions, resulting in either an increase in the size of lion prides or the establishment of new prides. This mechanism where an increase in one species can cause a decline in another mediated through a shared predator is called "apparent competition," because superficially the negative correlation between the abundance of the two prey species suggests a competitive interaction. Nevertheless, the question remained whether elevated predation risk could have been responsible for the more widespread population declines shown by sable and kudu.

Hence the following postulates or scientific hypotheses needed to be formalized and tested:

1 The direct effect of low annual rainfall affecting food production contributed to the declines.
2 The consequences of low dry season rainfall for green leaf retention during the dry season had an additional influence.
3 The effects of persistently low rainfall contributed to a degradation of the capacity of the vegetation to support these populations.
4 Elevated risk of predation following an increase in the abundance of lions, responding to increased availability of their main prey species, contributed to the population declines shown by less common prey species.

15.3.3 Formulating models for standard statistical tests

The next step is to evaluate the evidence provided by the data for these hypotheses. Note that they are not expressed as alternatives, because several factors could have contributed in combination to the population changes observed. Table 15.1 summarizes the data available for analysis for the focal antelope species. The original and derived data are also on the CD in spreadsheet format.

Statistical tests generally entail relating some response variable to a set of candidate predictors potentially affecting this response. Since we are interested in explaining population trends, the response variable is the change in population abundance, as estimated by the census total, between one year and the next. As population change takes place proportionately, it is best expressed using a logarithmic transform, that is, $\log_e(N_t/N_{t-1})$, where N_t represents the census total in year t. Since the predictor variables such as rainfall are also expressed as continuous measurements (rather than being categorized), this leads us to a multiple regression model. In general terms, such a model takes the form

$$Y = \beta_0 + \beta_1 X_1 + \beta_2 X_2 + \beta_3 X_3 + \cdots + \varepsilon,$$

where Y represents the response variable, the Xs the various predictors, and the βs the coefficients for the effect of each predictor. The factor ε is an error term for the variance unexplained by the set of factors included in the model. Interactions between factors, expressed as the products of specific combinations of variables, could additionally be incorporated.

Table 15.1 Annual census totals for three antelope species in the northern half of KNP together with associated rainfall records for this region (expressed relative to long-term mean values), plus derived indices of the effect of the previous 4 years rainfall on habitat conditions, and prior prey availability on the abundance of predators

Year	Kudu	Sable	Roan	Annual rainfall	Wet season rain	Dry season rain	Previous rain (4 years)	Prior predator food (4–8 years)
1977	2562	805	287	1.403	1.648	0.959	1.125	
1978	2689	882	277	1.399	1.333	1.009	1.355	0.973
1979	3033	993	295	0.895	0.892	1.293	1.297	1
1980	3485	1050	319	1.203	1.244	1.112	1.207	0.967
1981	4237	1213	323	1.528	1.469	1.876	1.225	0.852
1982	3862	1354	311	0.755	0.637	1.183	1.256	0.949
1983	3020	1391	332	0.464	0.515	0.980	1.095	1.059
1984	3324	1350	324	0.684	0.720	0.508	0.987	1.114
1985	3832	1476	352	1.430	1.303	1.950	0.858	1.154
1986	3588	1411	377	0.862	0.576	1.319	0.833	1.31
1987	2895	1322	233	0.430	0.483	1.679	0.860	1.516
1988	2600	1216	169	1.230	1.185	1.343	0.852	1.752
1989	2214	1128	175	0.609	0.660	0.472	0.988	1.905
1990	1805	1011	155	1.012	1.151	0.278	0.783	2.011
1991	1026	789	119	0.881	0.985	0.364	0.820	2.136
1992	811	568	58	0.410	0.399	0.375	0.933	2.109
1993	634	334	45	0.897	0.987	0.480	0.728	1.973
1994	605	308	45	0.729	0.814	0.611	0.800	1.875
1995	624	286	30	1.004	0.918	0.448	0.729	2.176

Alternative models differ in the number of factors included and the specific factors incorporated, as well as in the functional relationships that are assumed. Adopting a conservative approach, models could be restricted to factors that were measured directly, that is, rainfall or its seasonal components, and the population abundance level. Note that the latter is automatically included because basic ecological theory suggests that the population density is likely to modify the effect of any other factor, even weather variation, particularly if the latter operates through food availability. This leads to a comparison between just two models:

$$\log_e(N_t/N_{t-1}) = \beta_0 + \beta_1 N_{t-1} + \beta_2 R_a + \cdots + \varepsilon \qquad (15.1)$$

and

$$\log_e(N_t/N_{t-1}) = \beta_0 + \beta_1 N_{t-1} + \beta_2 R_w + \beta_3 R_d + \cdots + \varepsilon, \qquad (15.2)$$

where R_a represents annual rainfall, R_w wet season rainfall, and R_d dry season rainfall. Note that the form of density dependence incorporated into this model follows the Ricker, or exponential, logistic equation (see Chapter 3). Furthermore, rainfall variation is expressed using a logarithmic transform. This is done because a reduction in rainfall by the same amount below the mean is expected to have a greater effect on population growth than the same increase above the mean. The log transform means that we consider *proportional* variation in rainfall to be more meaningful (and this assumption could be tested by comparing models relating population growth to alternative measures of rainfall). A further assumption in specifying the model in this way is that we expect the effect of rainfall or its seasonal components to be additive, that is, the population growth rate will be higher for the same density level if the rainfall is higher. Moreover, the contribution of dry season rainfall to population growth is independent of that of the wet season component. Recall that in formulating the kudu population model in Chapter 6, we had assumed that rainfall and density acted in combination by using the ratio "rainfall/population biomass" as the predictor for the survival functions. The reason why we do not now consider this interactive influence will become evident below.

When the statistical support for the contribution of each predictor in these alternative models is compared, using the standard output from any statistical package, the second model above seems more strongly supported in terms of the R^2 value (Table 15.2). However, it appears that, while dry season rainfall is most influential for sable, for kudu the wet season component, which contributes mainly to the annual total, has the most consistent effect. For roan, both rainfall components seem to make similar contributions, as judged by the slope coefficients, but the statistical support indicated by the P values was very weak. Since the roan population was the smallest, sampling error in the count totals is relatively greater. Furthermore, the sign of the slope coefficient for the effect of the population abundance was *positive* for both roan and sable, contrary to what was expected. This means that, rather than manifesting a density feedback, the populations of these two species declined faster as they became less abundant. It seems that some spurious influence affected the density relationship. For kudu, abundance level showed almost zero effect. Accordingly, the relative population abundance was excluded as a factor in the model when assessing the influence of rainfall components for all three species.

However, the model incorporating the effects of both wet and dry season rainfall accounted for less than half of the variability in the data, for all three species (Table 15.2). This could be due merely to inevitable errors in the aerial census totals, as a result of the variable proportion of animals missed. Nevertheless, we need to consider whether a lagged influence

Table 15.2 Statistical output comparing two basic models relating annual population growth of three antelope species in KNP to rainfall. Model (1) includes the annual rainfall total as the sole predictor and model (2) the additive contributions of the wet season and dry season rainfall components

Species	Model 1			Model 2				
	Annual rainfall coefficient	P	R^2	Wet season rainfall coefficient	P	Dryseason rainfall coefficient	P	R^2
Roan	0.221 ± 0.138	0.128	0.139	0.190 ± 0.136	0.182	0.116 ± 0.084	0.189	0.235
Sable	0.145 ± 0.101	0.169	0.115	0.064 ± 0.086	0.471	0.173 ± 0.054	0.006	0.447
Kudu	0.247 ± 0.103	0.029	0.266	0.147 ± 0.063	0.033	0.167 ± 0.101	0.121	0.394

from either habitat deterioration or elevated predation risk, as expressed in postulates (3) and (4), could have contributed additionally to the observed population trends. The problem is that data on changes in vegetation and predators are not available.

The way forward is to derive proxy measures expected to have some relation with the underlying causes. For example, if persistently low rainfall results in vegetation degradation, we could relate population changes to the annual rainfall averaged over several previous years. Furthermore, if elevated prey abundance could bring about an increase in the lion population, we could relate the population changes of prey species to the earlier food abundance for lions, as indexed by the combined biomass, or more specifically the carcass biomass produced, of all prey species. This leads to two further models:

$$\log_e(N_t/N_{t-1}) = \beta_0 + \beta_1 N_{t-1} + \beta_2 R_w + \beta_3 R_d + \beta_4 R_p + \cdots + \varepsilon \quad (15.3)$$

and

$$\log_e(N_t/N_{t-1}) = \beta_0 + \beta_1 N_{t-1} + \beta_2 R_w + \beta_3 R_d + \beta_4 P_p + \cdots + \varepsilon, \quad (15.4)$$

each now potentially incorporating four predictors, with R_p representing prior rainfall and P_p past prey abundance for lions. Note that the density influence is still potentially included because the sign of its apparent effect could change when an additional factor is incorporated. Furthermore, the effect of prior rainfall and prior prey abundance are considered in separate models because they are not independent over time (i.e., closely cross-correlated in their variation). Specifically, the period after 1986 when prior rainfall had been consistently low was also the time when past prey abundance, as constituted especially by zebra, wildebeest, and buffalo, had been high. Note further that both of these predictors are hypothesized to have cumulative lagged effects, that is, their influence on population growth depends on conditions that prevailed over some period in the past.

The statistical output from the regression model indicates that prior rainfall had a significant influence on population growth only for sable, while the effect of past prey availability appeared significant for both roan and sable, and almost significant for kudu (Table 15.3). However, the influence of dry season rainfall disappeared when the prey availability index was incorporated. This is because the period after 1986 when dry season rainfall had been consistently low was also associated with high past prey availability and low prior rainfall. Hence adding any one of these factors to the model automatically reduced the effect of the others because they did not vary independently. How do we decide which of these three factors had been most influential?

Table 15.3 Statistical output for two additional models relating annual population growth of three antelope species in KNP to rainfall plus an additional lagged predictor, in addition to the effects of the seasonal rainfall components. Model (3) incorporates prior rainfall and model (4) past prey availability for lions, as the lagged predictor

Species	Lagged predictor	Wet season rainfall		Dry season rainfall		Prior rainfall or prey availability		R^2
		Coefficient	P	Coefficient	P	Coefficient	P	
Roan	Prior rainfall	0.184 ± 0.132	0.187	0.060 ± 0.092	0.527	0.375 ± 0.276	0.196	0.324
	Past prey availability	0.150 ± 0.116	0.219	−0.076 ± 0.102	0.531	−0.496 ± 0.189	0.020	0.487
Sable	Prior rainfall	0.057 ± 0.074	0.421	0.114 ± 0.051	0.043	0.396 ± 0.154	0.022	0.625
	Past prey availability	0.038 ± 0.074	0.614	0.050 ± 0.065	0.451	−0.318 ± 0.120	0.019	0.633
Kudu	Prior rainfall	0.163 ± 0.101	0.128	0.112 ± 0.070	0.131	0.234 ± 0.210	0.284	0.443
	Past prey availability	0.144 ± 0.096	0.154	0.041 ± 0.084	0.636	−0.275 ± 0.156	0.099	0.504

15.3.4 Model selection statistics using AIC

The model selection approach is aimed at identifying the most parsimonious model consistent with the data from among a set of candidate models, taking into account the dilution effect of including additional predictors that we noted above. Models (3) and (4) above represent alternative "saturated" models, including all of the factors that we recognized as being potentially influential (but recognizing habitat degradation and predation as alternatives because of their contemporaneous variation). Using this approach, we can additionally consider further models leaving out one or more of the potential predictors in different combinations.

This results in a set of seven candidate models, listed in Table 15.4. Four statistical measures are reported for each model, derived from model fitting by least-squares regression. The coefficient of determination, or R^2 value, indicates the proportion of the total variability explained by the model, as revealed by how small the residual sum of squares is relative to the total sum of squares. This is important because the best model may still explain very little of the variability in the data, particularly if the sampling error is great. The AIC values (corrected for small sample size as described in Section 15.2) are shown, although they mean very little directly. From these values, the model with the lowest (most negative) AIC_c was identified as the best-supported model, and AIC differences (ΔAIC_c) were derived giving the best model a value of 0. The relative likelihood of each model compared with the best model, the latter assigned a likelihood of 1.0, was then derived from their relative contributions to the total AIC difference summed across all seven of the models considered, as described in Section 15.2.

For all three antelope species, a model including the index of predation as a predictor was judged best. Indeed, for both roan and sable, the model incorporating predation alone appeared best because the variation explained by adding annual or seasonal rainfall did not counteract the penalty for the additional factor. Nevertheless, models including rainfall components as well as predation were 0.31–0.58 as likely as the model considering only predation, indicating that the effects of rainfall could not be totally discounted. Another data set, or a greater sample of years, could quite likely change the ranking among these models. For roan and sable, habitat change alone or in combination with seasonal rain appeared substantially less likely than predation as a factor causing the declines, but with relatively likelihoods of around 0.05 could not be completely discounted. For kudu, which provided the most robust counts because its population total was largest, the best model incorporated both the effects of predation and annual rainfall, although the model including predation alone was a close second. The additional variation seemingly explained by

Table 15.4 Model selection statistics for the dynamics of the three antelope populations in northern KNP

Model	Roan				Sable				Kudu			
	R^2	AIC_c	ΔAIC	Relative likelihood	R^2	AIC_c	ΔAIC	Relative likelihood	R^2	AIC_c	ΔAIC	Relative likelihood
WSR + DSR	0.235	−51	7.5	0.024	0.447	−67.3	9.2	0.01	0.394	−61.7	4.5	0.105
WSR + DSR + HAB	0.324	−49.9	8.6	0.013	0.625	−71	5.5	0.064	0.443	−59.8	6.4	0.041
WSR + DSR + PRED	0.487	−54.9	3.6	0.165	0.633	−71.3	5.2	0.074	0.504	−62.2	4.3	0.135
ANR + PRED	0.462	−57.4	1.1	0.577	0.621	−74.2	2.3	0.317	0.528	−66.2	0	1
DSRA + PRED	0.427	−56.2	2.3	0.317	0.626	−74.4	2.1	0.35	0.424	−62.6	3.6	0.165
PRED	0.408	−58.5	0	1	0.61	−76.5	0	1	0.415	−65.2	1	0.607
HAB	0.197	−53.1	5.4	0.067	0.461	−70.7	5.8	0.055	0.215	−59.7	6.5	0.039

Notes: WSR, wet season rain; DSR, dry season rain; ANR, annual rainfall; HAB, previous rainfall; PRED, prior food abundance for predators.

dry season rainfall was insufficient to discount the penalty for including it as an additional factor, or in place of annual rainfall, although based on relative likelihood its contribution should not be totally discounted. A negative density feedback was also indicated for kudu once the effects of both predation and rainfall were incorporated in the model. The full data set including regional distinctions in census totals indicated significant contributions from both wet and dry season rainfall, in addition to predation and population abundance, on the dynamics of kudu (Owen-Smith and Mills 2006). Furthermore, the effect of predation on kudu appeared strongest in the northern section of KNP, where most of the roan and sable occurred.

The model selection process indicates how alternative models are not simply rejected, as in null hypothesis testing. Instead, the relative weight of the evidence in favor of each model, provided by the available data, is evaluated, recognizing that another data set could support a different interpretation through chance sampling effects. We can now state confidently that an increase in predation pressure, putatively resulting from elevated prey availability, most probably contributed to the population declines shown by these three antelope species, without completely excluding the effects of current or prior rainfall on their susceptibility to predation. With data spanning only 18 years, it is basically unlikely that more than two or three factors could be supported by AIC, without taking into account the regional breakdown possible for kudu. Additional support for the effect of elevated lion abundance came from a separate analysis showing that variation in the estimated survival rates of adults of these antelope species was also correlated with past changes in prey availability for lions, while no such effect on juvenile recruitment was detected (Owen-Smith et al. 2005). It is important to recognize that the assessment is limited to the set of candidate models included. There could be some other as yet unidentified factor that was even more influential. Note also that the model with the highest R^2 value is generally not supported as the best model because of the penalty for including additional predictors.

15.3.5   Implications for conservation

Superficially, findings might suggest a justification for reducing the predator population. However, this is likely to be ineffective in the long term, as long as the prey base remains. The more fundamental question is, what was responsible for the increase in the food supply for the lion population?

Some of the increase in prey availability was due to an expansion in the zebra population, from a count total of around 16,000 in 1978 to over 30,000 after 1983 (Ogutu and Owen-Smith 2003). While the buffalo population did not increase, these animals featured more strongly in the prey of lions under the low rainfall conditions that prevailed after 1983 (Mills et al. 1995). Giraffe also increased in abundance, while the population levels of the less common ungulate species peaked during the early 1980s after rainfall had been consistently above average for several years (Fig. 15.1). All of these population changes contributed to a two-fold increase in the indexed food availability for lions between the mid-1970s and mid-1980s (Owen-Smith and Mills 2006). This would be sufficient to produce a doubling in mortality due to predation, all else being equal.

A more fundamental factor underlies the increase in the abundance of zebra and other water-dependent species: the establishment of numerous water points in the form of boreholes and dams augmenting natural water sources throughout KNP, aimed at reducing the vulnerability of ungulate populations to droughts (Grant et al. 2002). This change in the ecosystem context evidently benefited the more common ungulate species, which form the main prey base for lions, at the expense of less common species (Owen-Smith and Mills 2006). Recognizing this, park managers have instituted a programme to reduce the numbers of artificial water sources. Whether this will be effective in enabling the recovery of the rarer antelope species remains to be seen.

15.4 Overview

This chapter outlined how a wide-ranging statistical approach based on information theory can be used to evaluate the support provided by the available data for alternative explanatory models. The example considered was perhaps unusual in the amount of data available, although some potentially crucial information was missing. Rather than ignoring these gaps, proxy measures were derived to fill them. With less information, or greater sampling errors in the estimates, there would be less discrimination among the alternative models.

You should have learned the following:

1 Models differ in the explanatory factors included and the functional form assumed for their effects.
2 Models are compared in terms of the amount of information in the data that they explain, using an information theory measure such as the AIC.
3 These measures balance the amount of unexplained variation in the data, indicated by the residual sum of squares or deviance, against the diluting effect of additional predictors.

4 The procedure identifies the most parsimonious model consistent with the data, in terms of the number of factors included, as well as relative support for alternative models.
5 A larger data set is likely to provide support for more factors through containing more information.
6 The coefficient of determination indicates how much of the variation in the data is explained by the best model.
7 The statistical associations revealed by the model fitting may be used to infer the processes most likely to be responsible for declining population trends.

Recommended supporting reading

The book by Burnham and Anderson (2002) provides a comprehensive account of model selection statistics. More concise guidance may be obtained from the papers by Anderson, Burnham, and Thompson (2000), Johnson and Omland (2004), and Hobbs and Hilborn (2006). Hilborn and Mangel (1997) outline the philosophical perspective underlying statistical diagnosis. An application to choosing the equation best fitting a data set is described by Sinclair et al. (2005, Chapter 15). Pain and Donald (2002) consider the extent to which population declines by birds might be due to climate change versus human impacts such as agricultural intensification, without invoking formal statistical procedures. Further information on changing ungulate populations in the KNP is provided by Owen-Smith and Ogutu (2003).

Programs on the accompanying CD

An Excel file containing the data for the three antelope populations and information on rainfall, plus illustrations of the multiple regression analysis in Excel, is on the CD.

Exercise

Apply model selection statistics to any data set available to you, or obtained from the literature.

Appendices

Appendix 2.1 Guidelines for the dodo population model

The simplest model to use for the dodo population is an accounting model, that is, the population size at some future time is equal to the population size now, plus births and immigration, less deaths and emigration. This is commonly labeled with the acronym BIDE: Births, Immigration, Deaths, Emigration. Since emigration and immigration are precluded in the island situation, you can simplify its formulation even further to represent just births and deaths. Hence, mathematically,

$$N_{t+1} = N_t + B - D,$$

where N_{t+1} represents the number of dodos at some later time $t + 1$, N_t represents the number of dodos now, B represents births or, in this case, dodos hatched, and D represents deaths. Since breeding is likely to be seasonal, it would seem best to increment time in units of 1 year.

You know that the starting population at time zero, that is, N_0, is two dodos. How many births will occur during the first year? Obviously zero because the population consists of two newly hatched chicks, too young to breed. This raises a leading question, at what age do dodos first reproduce? Since this information is not available, you will have to make a guess. On what basis could your estimate be an informed guess rather than a wild one? Think about this. I will offer a suggestion in Appendix 2.2 that covers the assumptions that you might have made.

The total population size will remain at two (assuming no mortality), until the founder pair reaches the stage at which they have mated, laid the eggs, and hatched the first offspring. You will need to make further assumptions to calculate the annual number of births thereafter, in order to keep track of how the population size increases over time. Since the starting population consisted of just two individuals of identical age, recruitment of new hatchlings into the population will take place erratically for several years. You might have considered using an exponential growth equation, but this would be inappropriate because the age and sex structure of the population affects the increment of new birds each year. It would be more accurate to keep account of the survival and reproductive output of each individual bird.

Ultimately the population size will be limited by the total area available to accommodate breeding territories. The data provided in Box 2.1 give some indication of the size of these territories. A logistic growth equation would not be the best way to represent this population ceiling, because it is unlikely that there would be any reduction in the reproductive performance of the birds until the density limit set by territory area was reached. On the CD you will find an example of an elementary population model that might have been appropriate, labeled ELEMPOP. Try to understand the file written as text in the programming language of True BASIC. Some explanation can be found in the text files accompanying the programs on the CD. Run the model ELEMPOP to see what information it needs and what output it generates.

Appendix 2.2 Assumptions that you might have made

Besides the assumptions that you recognized, there were probably other implicit assumptions that you made in formulating the model, without thinking about them. Below is a list of some that came to my mind.

1 What basis did you use for estimating the age at first reproduction and the interval between successive breeding (i.e., batches of eggs laid)? Lacking specific information for dodos, the most defendable assumption would be to use the values recorded for similar-sized birds. This assumes that vital rates depend fundamentally on body size. What other aspects of evolutionary history and environmental features might influence reproductive rates? It is doubtful that birds as large as dodos occupying islands with restricted space would reproduce annually.

2 What did you assume about the sex of the two dodo embryos? What is the possibility that the two eggs being incubated would turn out to be both male, or both female? Can you validly assume that the sex ratio thereafter would be exactly one male to one female for each batch of eggs hatched? Chance effects like random deviations in the birth sex ratio can make a big difference to the population growth when the population size is small.

3 Did you assume that the breeding interval remained fixed despite the likelihood that fruit production would vary substantially from year to year?

4 Did you assume that the entire island would be a suitable habitat for breeding territories? What about the adequacy of the food resources,

recognizing that waterberry trees and other fruit-producing plants are unlikely to be evenly distributed over the island?

5 What shape did you assume for the territories? If circular, did you simply divide the circular area into the size of the island without considering the vacant space that exists between contiguous circles?

6 Where did you assume that young birds went after leaving their parents? If tolerated by territory holders, they form part of the population, even though they contribute nothing to recruitment. In this situation, territorial space limits the size of the breeding population, but not the total population.

7 Did you consider providing supplementary food to ensure that the initial hatchlings and their later offspring thrived? What might be the cost of providing this food?

8 Did you give any consideration to genetic factors and how they might influence the viability of the population? What genetic aspects need to be considered? How could genetic problems be overcome, and at what cost?

9 What other factors did you consider in assessing the risk that the whole enterprise would fail? Did you take into account possible accidents, environmental hazards, or the chance of a catastrophe such as a cyclone occurring?

10 How could the chance of complete failure be reduced? What would be the additional cost? What role might *ex situ* conservation in zoos play in aiding the conservation of the species.

If you won the competition and were offered the chance to study the population, what extra biological information would you want to gather? How would you go about doing this study without threatening the success of the conservation operation?

Appendix 2.3   Overview of what you should have learned

This chapter introduced a hypothetical conservation biology problem with typical features: missing information, uncertainty in the information that is available, risk of failure, and financial constraints. Hence, assumptions had to be made in developing the model. Some of these may be explicitly recognized, while others were covert – you were not conscious of these assumptions until later reflection.

You should have learned further that incomplete information does not stop you from building a model. In fact, the less complete the information, the more you need a clearly specified model to help expose the various assumptions made in projecting future scenarios. The specific answers that you got to the questions posed are of lesser importance than the logic and approximations that you used to get these answers. It is the latter that needs to be defended when you find that your answers differ from those of other teams working on the same problem, or alternatively abandoned if you concede that other teams had in fact been more thoughtful. The basic question is, how much confidence do we have in the answer that was provided? When information is unreliable, it can be helpful to consider a range of possible estimates, from "conservative" to "optimistic," rather than just one "best guess."

Among the range of models that could have been adopted, it is generally best to choose the simplest model that is adequate for its purpose. Why add density dependence when there is no information on how it might occur, apart from the potential limit on the population size set by territory size (assuming this is inflexible)? However, adopting a simple exponential growth equation would be misleading because the starting population structure does retard the initial increase rate. Moreover, with such a small founding population, chance effects need to be considered, for example, the possibility that not only the initial hatchlings, but also their subsequent offspring could be of the same sex. Lack of genetic variability could also become a constraint on population performance in such a situation.

Each additional level of complexity in the model requires more information, and the information available is commonly rather limited. While a computer would be helpful in performing the calculations needed, both for the population dynamics and for the financing, the facilities that it brings could have tempted you into formulating a structurally more elaborate model than is justified. A useful guiding principle is expressed by the acronym KISS – Keep It Simple, Stupid! In other words, do not use a more complex model than is justified by the available data and knowledge.

Appendix 3.1   Exponential growth model formulated in a spreadsheet

Exponential growth basically involves growth by a constant proportion, that is, $N_{t+1} = N_t + rN_t$, where r is the proportional growth rate. Hence in your worksheet, type EXPONENTIAL GROWTH at the top, and underneath it write out the formula for exponential growth, as text. Leave a few blank lines, then set up the following column headings: "TIME," "POPULATION,"

"INCREASE," and "PROPORTION." Above these headings, in column A type "$r =$" then in column B type a suitable annual value for r, say 0.25. Increment time from 0 to 50 years, then at time 0 enter a starting number for the population size, say 2, in column B. In the same line in column C, enter a formula to calculate the value of the annual population increase, which is simply the population size times its growth rate r. Remember to anchor the value of r in the cell where you have typed it using the "$" sign. In Column D, enter the proportional rate of population growth, which is the numerical increase divided by the population size at that time. The formula needed in the next row of column B for the population size one time step later is simply the previous population size plus the annual population increase. Copy these formulae down over the 50-year period. The last line in column B will indicate how large the population has become after 50 years.

Last, generate a graph showing how the total population size changes over time, as a simple line chart. Then transform the Y-axis to a log scale. Finally, plot the proportional growth rate against the population size. This will need to be an XY (scatter) plot (otherwise the sequence of numbers on the X-axis will be strange) but with a line joining the points in the plot.

Strictly this is geometric rather than exponential growth because you repeated the calculations at a discrete time interval of 1 year. True exponential growth takes place continually during the course of the year. You can approximate this by shortening the time step, for example, to monthly. Hence insert a new column B headed "YEARS." It gets incremented by one unit after every 12 time steps. Next divide the value that you had previously entered for the annual population growth rate r by 12, the number of months in a year. To obtain the population size after 10 years, you will need to extend the number of rows to span 120 time steps. You should find that the population size after 10 years is now slightly larger than that projected when you used annual time steps. This is because individuals added to the population during the course of the year contribute to further population growth.

Reduce the time step to a week and see how much further difference this makes to the final population size after 10 years. Don't forget to adjust the population growth rate after altering the time step. Then use a daily time increment and view the population size 3,650 lines down. Check your answer against that produced by true exponential growth by entering the equation $N_t = N_0 e^{rt}$ as a spreadsheet formula in a blank cell of the worksheet. Set $t = 10$ years and $r = 0.25 \, \text{year}^{-1}$. The smaller the time step you use for geometric growth, the closer the answer approximates true exponential growth.

Appendix 3.2   Converting the geometric growth model into a ceiling model

For the ceiling model, the IF function provided by the software needs to be used to ensure that the population size does not exceed some maximum level. The basic logic is that IF a certain condition is TRUE, one operation follows, ELSE an alternative action occurs. Hence modify the worksheet that you created for exponential growth as follows:

1 Above the headings in column C enter the words "MaxPop =," and then in column D the value for the maximum population size, say 200.
2 In the row for time 1 in column B, change the formula to read

$$= \text{IF}(\text{B5} + \text{C5} > \text{D\$3}, \text{D\$3}, \text{B5} + \text{C5}),$$

where the row numbers may need to be adjusted to match the appropriate rows in your worksheet. This conditional formula means that IF the value for the population size calculated by adding the population increase to the previous population size is greater than the value specified for the maximum population size, the population size becomes equal to this maximum, ELSE it takes on the value given by the addition.
3 Extend all of the calculations down to sufficient rows to allow the population to reaches its maximum size. The population grows exponentially until it reaches the ceiling density, at which level further increase is abruptly halted. Note the changes to the graphs, both for population size versus time and for the proportional growth rate versus the population size. You will find that you need to adjust the figures in the column for the proportional growth rate to zero once the ceiling level has been reached, and further growth thus prevented.

Appendix 3.3   Formulating the logistic model in a spreadsheet

The easiest way to start is by copying across the worksheet that you developed for the ceiling model into a new worksheet in your spreadsheet folder. Change the heading to read "LOGISTIC GROWTH." Also alter the series reference for the graphs to refer to the new sheet. In column C, change the formula for calculating the annual population increase ΔN from that for exponential growth to that for logistic growth. You should already have values for the two parameters, the maximum population growth rate r_0 and for the maximum population size K, entered in columns B and D above. Also remember to remove the conditional "= IF" statement from the formula for the population size in column B because it is no longer necessary.

Note in particular how both the annual population increase and the relative growth rate change as a function of the population size.

Appendix 3.4 Guidelines for developing a spreadsheet model for a set of descriptive equations

Develop each model in a separate worksheet in the same folder. Start by copying across the logistic model that you developed above into the first worksheet. Under the name of the model, enter corresponding equation, expressed algebraically as in the chapter text. Below the equation are the cells that contain the values of the fixed parameters (i.e., r and K for the logistic equation). Leave about 15 blank lines, then move the graphs depicting the model output into this space. Accordingly, starting at about row 22, enter the following headings for the columns that will contain the numerical data: Time; Population; Increase; Prop. Rate.

Increase time from 0 to 100 to span a century. Start the population at time zero with some small number, say one or two animals. Enter the appropriate equation (in spreadsheet form) to calculate the annual population increase (the amount by which the population size changes each year) generated by this population, and then add that to the current population to get the population size the next year. Note the logic: the current density generates a certain population growth, which causes the population to increase (or decline) between this year and the next. To obtain the proportional (or relative) growth rate, divide the annual increase by the population size in this year. Copy these formulae down across the time range from 0 to 100 years. Don't forget to anchor the cell references for the fixed parameters in the cells that contain these values. Finally generate the appropriate graphs. Only two are needed. One should show the trend in total population size over time. The other should depict the dependence of the annual increase and proportional growth rate on the population size. Plot the increase on the secondary Y-axis because of the different ranges in numerical values that these two plots represent. Make sure that you have selected an XY chart rather than a line chart for this graph.

For each new model, copy across the worksheet you developed for the previous model, then change the equations under the name and in the column where the basic calculation is done. Note that for the Ricker and Beverton–Holt models, the equation is used to calculate the population size one time step later, rather than the annual increase as was done for the logistic model. The increase is then obtained by subtracting the population size in the same line from the population size reached in the next line. Be aware that the Beverton–Holt equation uses the population growth factor λ (i.e., $1 + r$), and incorporates curve-fitting parameters a and b rather than specifying the carrying capacity directly (the latter is inherent in the values for a and b, as explained in the text). Also, the entire term aN_t should be enclosed in brackets, and raised to the power b.

In the threshold logistic model, the population growth rate remains fixed at its maximum until the threshold density is surpassed, and only thereafter is the effect of density on this growth rate taken into account. Refer to Appendix 3.2 for guidance on how to specify this conditional outcome using the IF function. Also note that the threshold value of the population size at which density dependence takes effect must be subtracted both from the population size and from the value of K in the formula for calculating the annual population increase.

For the lagged logistic model, the formula for density dependence is adjusted simply by replacing N_t by $N_{t-\text{lag}}$, where lag represents how many years back in time the effective density lies. Obviously this formula cannot be applied when fewer years have passed since the start of the sequence of years than the specified lag time. Hence perform the calculations ignoring the lag for these initial lines.

For the models incorporating random variation in K, you will need an extra column, which can be labeled "Environment." Allow the environment to vary evenly between 0.5 and 1.5 times its mean by entering the random number function "$= \text{RAND}()+0.5$" in the first line, and copy it down. In the population growth equation, multiply the basic value of K specified at the top of the worksheet by the state-of-the-environment factor for each year. With the annual population increase and rate varying randomly from year to year, lines connecting successive values in the density-dependent plots will look rather wild and woolly. Hence in these graphs replace the lines by unconnected symbols to produce scatter plots. To confirm that the same general trend in density-dependent patterns persists, fit curved trend-lines through these sets of points. Note that each time you make any change to the worksheet, the sequence of random numbers changes, and so does the pattern of the population time series.

Appendix 4.1   Formulating an elementary age-structured population model in a spreadsheet

Set up the first few lines of the spreadsheet as you did for the previous chapter, with a heading at the top and space below to insert a graph spanning the next 12 lines as well as parameter values. Then, assuming the maximum longevity to be 6 years, type the following column headings: TIME; N0; N1; N2; N3; N4; N5; TOTN; INCREASE; PROPRATE. Assume further that the population is censused immediately after births have occurred, so that N0 represents the number of newborn animals aged exactly 0 years, N1 the number aged 1 year, and so on. This makes accounting for offspring mortality between birth and the time of the population count unnecessary. In the first row after the column headings, set the starting time to zero, and enter the initial number of animals in each age class. Use the "sum" function in the spreadsheet to calculate the total population size, and then enter the formulae to calculate the annual population increase and proportional growth rate as you did in the spreadsheet models developed for Chapter 3. In two rows immediately above the column headings, enter parameter values representing the age-specific fecundity and survival rates, respectively. Convenient values would be a fecundity of zero for age class 0 and 1 for the remaining age classes. The annual survival rate is typically low for the first age class, say 0.5, and then 0.9 for each older class. Remember to make the survival rate for the last age class zero because we have assumed that individuals do not live beyond 6 years of age. Since by convention the model represents only the female segment, a fecundity of 1 means that each female produces two offspring annually, of which exactly half are female.

Recall that individuals in each age class come from the number of animals that were a year younger the previous time, multiplied by their survival rate between this age class and the next. The survival rates at the top of the table refer to survival *from* the age class in this column to the next age. This is why the survival value for the last age class is zero, that is, no animals aged 5 years survive to become over 6 years of age. All age classes contribute to newborn animals (i.e., N0) the next year, provided they have nonzero fecundity. However, the contribution of each class must be multiplied by the proportion of these females that survive to the next time step, since those failing to survive cannot contribute offspring. If the population had been enumerated some time after the birth pulse, you would need to consider additionally the possibility that some of the offspring born may not survive until the census period.

Copy the formulae down over a period sufficient for the relative growth rate to stabilize, say 25 years, making sure that fixed parameters are anchored in the appropriate cells. With the combination of vital rates suggested above, you should observe an eventual growth rate of just over 14% per year. Graph both the change in population size over time and the change in relative growth rate with time. Note how the annual growth rate fluctuates initially before eventually stabilizing. Experiment by changing the initial age composition, for example, starting with a population consisting entirely of newborn animals in age class N0, or just of adults in age class N4 (why not N5?). The same relative growth rate is always reached, whatever the starting age structure. This is the inherent population growth yielded by the specific set of vital rates. Also calculate the proportion of the total population made up by different age classes when the overall population growth rate stabilizes. These proportions should also end up the same, and represent the stable age distribution for these vital rates.

Having set up the model with some basic parameter values, undertake a sensitivity analysis by investigating how much difference it would make to the eventual population growth rate if certain vital rates were altered. For example, what if the survival rate over the first year was lowered from 0.5 to 0.4? What if the survival rate of all adults, or of just one adult class, was lowered by the same amount, from 0.9 to 0.8? What if the fecundity was lowered from 1 to 0.9? More generally, which vital rate most sensitively influences the overall population growth rate?

However, it might be more meaningful to adjust each vital rate by the same proportion, for example, lowering first-year survival by 10% from 0.5 to 0.45, adult survival from 0.9 to 0.81, and fecundity from 1.0 to 0.9. The assessment of the relative effects of *proportional* adjustments in vital rates is termed an *elasticity analysis*.

Also investigate the effect of increasing the longevity of the animals beyond 6 years. Having done this, collapse the adult age classes into one single stage with a constant survival rate in this stage, and note how much difference this makes to the projected population growth rate. The model representing persistent survival in the final adult stage is equivalent to an Usher matrix model.

Appendix 4.2   Formulating a stage-structured model for a sea turtle population in a spreadsheet

The loggerhead turtle is a long-lived marine reptile that occurs widely in different oceans. The animals are difficult to age due to fast juvenile growth, while their brittle shell does not readily retain tags. They are under threat due to human harvests of eggs and hatchlings on the beaches where they nest, and through adults becoming entangled in fishing nets. The data for this exercise come from a study by Crowder et al. (*Ecological Applications* 4: 437–45, 1994). These authors distinguished five life history stages: (i) hatchlings, (ii) small juveniles, (iii) large juveniles, (iv) subadults, and (v) adults. Due to the indeterminate growth of turtles, allowance must be made for the proportion of individuals that remain in the same stage 1 year later, as well as for the annual proportion growing into the next stage. The overall survival rate is the sum of the persistence rate and the transition rate for each stage.

Below is a table showing the vital rates estimated in this study, arranged in the form of a stage transition matrix:

	Hatchlings	Small juvenile	Large juvenile	Subadults	Adults
Hatchlings	0	0	0	4.7	62
Small juvenile	0.68	0.70	0	0	0
Large juvenile	0	0.05	0.66	0	0
Subadults	0	0	0.02	0.68	0
Adults	0	0	0	0.06	0.81

The numbers in the first row represent the effective fecundity, that is, the number of female hatchlings contributed by each individual female annually. The other entries are the persistence rates within a stage and transition rates between stages (which is which?).

Set up your worksheet with columns to represent time in annual steps, the five stage classes, total population size, annual increase, and proportional growth rate. Three rows at the top are needed for the stage-specific vital rates. The first row contains the fecundity rates. The second row contains the transition rates between stages. The third row contains the persistence rates within stages.

Based on these rates, work out the intrinsic growth rate for this turtle population. Start with the following initial numbers in each stage class: 300 hatchlings, 500 small juveniles, 180 large juveniles, 200 subadults, and 1000 adults. What difference does it make if you alter these proportions?

Recognizing that the species is threatened, consider how best the population growth rate could be enhanced by conducting a sensitivity or elasticity analysis on the vital rates. In particular, evaluate these alternative options:

1 incorporating "turtle exclusion devices" into fishing nets, which increase survival among large juveniles, subadults, and adults by about 10% per year, versus
2 protecting nesting beaches from human and other predators, so as to double the effective fecundity rate.

As survival is the sum of the persistence and transition rates, for option (i) you need to multiply *both* of the respective rates for each stage by a factor of 1.1 to represent a 10% improvement in overall survival. Based on your findings, what recommendation would you make?

Having developed this stage model for the turtle population, modify it to represent the population dynamics of a plant population. This entails adding additional vital rates to represent stage reversals as well as persistence in the seed stage.

Appendix 5.1   A simple interactive model for a herbivore–vegetation system

Start as for previous chapters by giving your spreadsheet a title in the first row. Several lines down, head four columns TIME, VEGET, HERBIV, and CONSUM. Assume that the vegetation growth follows a logistic model whereas the consumption of vegetation by the herbivores is linearly related to the amount of vegetation. Hence, just under the title type out the respective equations as text: $\Delta V/\Delta t = r_V V(1-V/K_V)$ and $\Delta H/\Delta t = caVH - mH$. In the next few lines, enter the values assigned to the parameters in these equations above the two column headings where they apply. Let $r_V = 0.5\,\text{year}^{-1}$ and $K_V = 500\,\text{g m}^{-2}$. Let $a = 0.02$, $c = 0.5$, and $m = 4.5\,\text{year}^{-1}$. Note that the annual gain in herbivore biomass over a year, equal to the product caV, amounts to five times the actual herbivore biomass when the vegetation biomass is at its maximum. This is counterbalanced by metabolic and mortality losses amounting to 4.5 times the herbivore biomass. Hence the herbivore population can potentially increase at a proportional rate of $5 - 4.5 = 0.5\,\text{year}^{-1}$.

Increment time in steps of 1 year, starting from time 0 over a period of 20 years. In the row for time 0, make the initial amount of vegetation equal to its saturation biomass of $500\,\text{g m}^{-2}$, and enter a very small herbivore biomass, say just $1\,\text{g m}^{-2}$. In the same row, type in a formula for the product aVH to calculate the amount of vegetation consumed over the course of each year. To start with, assume that this consumption has no effect on the amount of vegetation the next year. Hence in the vegetation column, enter the logistic equation to calculate the vegetation biomass in the next year: $V_{t+1} = V_t + V_t r_V(1 - V_t/K_V)$, as a spreadsheet formula. In the herbivore column, enter the formula to calculate the herbivore biomass the following year as a result of the amount of vegetation consumed, that is, $H_{t+1} = H_t + c\,(\text{consum}) - mH_t$, where the label "consum" refers to the amount of vegetation calculated from the formula aVH in the column headed CONSUM. Then copy these formulae down over the 20-year period. Last, generate a line graph showing how the vegetation and herbivore biomass change jointly over time. Note that these will need to be plotted on different Y-axes because of the large difference in their numerical values.

You should find that the herbivore population continues growing at a constant exponential rate, and eventually outweighs the vegetation biomass, which remains constant at its saturation level. This is obviously completely unrealistic.

The modification required is to subtract the amount of vegetation consumed by the herbivores from the total vegetation biomass. The vegetation then recovers at the rate governed by the logistic equation. In the formula calculating the vegetation biomass, replace V_t by $(V_t -$ consum) in both places where it occurs, where "consum" refers once more to amount appearing in the CONSUM column of year t. Due to regrowth, the vegetation amount the next year is reduced less than the total amount consumed by the herbivores over the course of the year. Copy this formula down, extending the time range to 100 years. Now you should see the growth of the herbivore population checked after it has reduced the vegetation biomass below $450\,\mathrm{g\,m^{-2}}$, using the suggested parameter values. After some initial oscillations, the herbivore biomass settles to a level around 1% of the vegetation biomass. Does this seem reasonably realistic?

To understand why this occurs, note that the herbivore consumption, and hence its population growth rate, is linearly related to the vegetation biomass. Hence a 10% reduction in the vegetation biomass, from 500 to $450\,\mathrm{g\,m^{-2}}$, means a 10% reduction in the biomass gained annually by the herbivore population, from 5 to 4.5 times the herbivore biomass. At this level the gain just matches the metabolic and mortality losses, which amount also to 4.5 times the herbivore biomass, based on the suggested parameter settings. Hence any further reduction in intake leads to a negative growth rate.

Explore the consequences of altering the values of each of the parameters in the model. Start by adjusting each by a very small amount. The output dynamics are highly sensitive to several of the parameters, and a large change is likely to result in an unrealistic descent into negative numbers, or failure by the herbivore population to increase at all.

Appendix 5.2 Incorporating a saturating functional response

It is unrealistic to expect the growth rate of the herbivore population to continue increasing, without limit, as the amount of vegetation increases.

Specifically, the amount of food consumed by the herbivores per day will eventually reach the upper limit that their intake rate and digestive capacity can maintain. This could be represented either by specifying a conditional limit for the linear intake function or by making the intake response a curvilinear function of food availability toward some upper asymptote. The latter is the modification made conventionally.

Copy the previous model into a new worksheet, and at the top change the text equation for the rate of growth of the herbivore population to read "$\Delta H/\Delta t = caVH/(b + V) - mH$." This means that a value is needed for the new parameter b, representing the amount of vegetation for which the intake rate is half of its maximum. Set $b = 100 \, \mathrm{g \, m}^{-2}$. The symbol a has a different meaning in the changed formula. It represents the upper limit to the amount of food that can be consumed over the course of a year relative to the herbivore biomass. Set $a = 10$. With these settings, you will need to reduce m to 4.0 to allow the herbivore population to grow. Replace the formula used to calculate consumption by the new intake function, appropriately anchored on the cells containing the parameter values. Copy the new formula down over the 100-year range.

With these parameter settings, you should observe dampened oscillations as you did before with the linear functional response. The initial growth rate of the herbivore population should be just under 0.17 year^{-1}, which is reasonably realistic, while the equilibrium biomass attained by the herbivores is a bit greater than with the previous model. As before, explore the sensitivity of the outcome to changes in particular parameters. You should find that the dynamics are somewhat less affected by changes to the vegetation parameters than before. Raising the saturation biomass of vegetation sufficiently still causes the system to crash, but only if the vegetation biomass gets reduced below half of its maximum setting. The output remains highly sensitive not only to the settings of a, c, and m, but also to that of b.

In addition to the plot of biomass changes over time, create a graph to show how the growth rate of the herbivore population changes as a function of the herbivore biomass density. This requires adding an extra column to the worksheet to calculate the annual proportional growth in herbivore biomass. You should observe a curve spiraling in toward the zero growth level, as a result of the oscillations. If you suppress the oscillations by adjusting one of the parameters slightly (e.g., raising m from 4.0 to 4.1), the spiraling disappears while the density response remains slightly convex. Note, however, that this has been achieved through slowing the maximum growth rate of the herbivore population.

To reveal the delay inherent in the density response, generate a duplicate chart plotting the annual growth rate of the herbivore population against the herbivore biomass density several years earlier. This should reveal an almost linear feedback response, once the appropriate lag time is selected.

Last, calculate and graph the herbivore population growth rate as a function of the vegetation biomass. This can be done directly using the equation for herbivore dynamics shown as text at the top of your worksheet, setting up two new columns. Note the curvilinear response to rising food availability, which follows directly from the form of the intake response. The suggested parameter values give zero vegetation growth at a vegetation biomass of $400 \, \text{g m}^{-2}$.

Appendix 5.3   Introducing competitive interference

Once again, copy the previous model into a new worksheet. To introduce an interference term, just two small changes are needed. First, the value for the interference coefficient d must be specified. Second, the formula for calculating consumption by the herbivore must be altered to become "$aVH/(b + V + dH)$." Retaining the same settings for other parameters as recommended above, you should find that a sufficiently large value for d dampens the population oscillations that would otherwise be manifested. Observe also the effect of increasing values of d on the form of the density-dependent response in herbivore growth rate.

Appendix 5.4   Allowing for environmental variability

The simplest way to introduce annual variability in the environment is to allow the saturation biomass level of the vegetation K_V to vary between years, presumably in response to variability in rainfall. Again copy the previous model into a new worksheet, and insert a new column indexing the state of the environment. Modify the number returned by the random number generator in this column by inserting the function $=(\textbf{RAND}()+0.5)^\wedge 0.5$, and copying it down. By taking the square root, the range of variation in the environment is compressed from 0.5–1.5 to 0.71–1.225, so that the dynamics do not become too wild. In the formula calculating the vegetation dynamics, replace the term representing K_V in the denominator by the cell reference to the basic value for K_V, multiplied by the state of the environment. Copy it down, and set the value for the interference term d to zero.

You should find that oscillations in the abundance of herbivores and vegetation, which previously were dampened for appropriate parameters settings, now tend to be recurrent. However, the pattern changes each time you alter the random number sequence by making a new entry in any cell of the worksheet, or either by pressing the function key F9. Observe also the form of the density dependence plot. This will appear confusing, unless you replace the lines connecting sequential points by a scatter plot of symbols. Can you discern the density feedback, either in this graph or in the next one showing the delayed density dependence? Increase the value of d above zero, and note how this changes the form of density dependence. Note further how the relationship between the herbivore growth rate and the amount of vegetation remains clearly defined even when some blurring is introduced by strong interference.

Appendix 5.5   Modeling the optimal stocking level for a livestock ranch

Title this new spreadsheet "SUSTAINABLE HERBIVORE STOCKING LEVEL," then specify the following equations for vegetation and herbivore biomass dynamics: $\Delta V/\Delta t = r_V V(1 - V/K_V)$, and $\Delta H/\Delta t = caH(V - v_u)/\{b + (V - v_u)\} - mH$. This means that we assume vegetation growth to be logistic, while the rate of increase in the herbivore population incorporates a hyperbolically saturating intake response. In the next few lines, enter these starting values for the parameters: saturation vegetation biomass $K_V = 5,000\,\mathrm{kg\,ha^{-1}}$; ungrazable vegetation biomass $v_u = 500\,\mathrm{kg\,ha^{-1}}$; maximum vegetation growth rate $= 0.5\,\mathrm{year^{-1}}$; maximum herbivore intake rate $a = 10$ times herbivore biomass per year; half-saturation level of vegetation $b = 400\,\mathrm{kg\,ha^{-1}}$; conversion coefficient from vegetation to herbivore biomass $c = 0.6$; metabolic herbivore biomass loss $m = 3.5$ times herbivore biomass per year.

Note the few small modifications from the previous model. First, the change in vegetation units from grams per square meter to kilograms per hectare, an adjustment by a factor of 10, has been done to correspond with the measure generally used in ranching contexts. An ungrazable amount of vegetation v_u is specified, to ensure that some vegetation remains to enable recovery, even if all accessible material were to be consumed by the herbivores. It is incorporated into the herbivore intake response. The value for the maximum annual food intake by the herbivore was obtained by noting that a cow typically consumes 2.5% of its body mass per day, which over the course of a 365-day year integrates to around 10 times the mass of the cow. The value for the half-saturation level of vegetation was derived from observations on cattle feeding in natural pastures. A lowered value for m has been assumed because mortality losses are negligible in the ranching context. Hence m merely represents metabolic attrition.

Next set up the following six column headings: TIME, VEGBIOM, VEGPROD, HERBIOM, HERBINT, and HERBPROD. Enter the appropriate spreadsheet formulae corresponding with the equations written at the top of the worksheet. Note that the annual vegetation production is calculated in a separate column, and then added to the vegetation biomass prevailing in that year, less the amount consumed by the herbivores, to get the vegetation biomass the following year. This enables you to keep track of the annual forage production by the vegetation resource as the herbivores graze it down. The herbivore biomass, rather than varying dynamically in response to changing vegetation resources, should be held fixed at the chosen stocking level. The annual intake or consumption by the herbivores is calculated using the formula for a hyperbolic intake function, as specified above with allowance for the ungrazable portion of the vegetation. The annual herbivore production is simply the intake multiplied by the conversion coefficient c, less the product of the metabolic loss m and the herbivore stocking biomass. Copy these formulae down over a time range of 30 years. This is a reasonable planning horizon for a rancher.

For consistency, the herbivore stocking level must be in the same units as the vegetation biomass, except that herbivores are conventionally assessed as live mass and vegetation as dry mass. In practice farmers normally assign stocking levels in terms of hectares per standard livestock unit (LSU), represented by a 454 kg cow. A low stocking level would be about 20 ha per LSU, equivalent to about 22 kg ha^{-1}, a high one 2 ha per LSU, amounting to 227 kg ha^{-1}. Experiment in the model starting with a low stocking density, then working upward. For each stocking level, establish the mean annual gain in herbivore biomass over the 30-year period. It might be helpful to include graphs showing the change in vegetation biomass, and in the annual herbivore production, over this time horizon. Establish the stocking level that gives the maximum averaged gain over this period, and note also the change in the vegetation biomass retained on the ranch at the end of the 30-year period. You should find quite a sharp threshold between the stocking level giving maximum yield, and one that causes the vegetation resource to crash to unproductive levels.

Note that the ranch was acquired in an ideal state, with the vegetation initially at its maximum potential. Now imagine the ranch being passed down as an inheritance to a son or a daughter, who gets it in the state left after 30 years. Does this make any difference to the productive capacity of the ranch over the next 30 years? A second graph showing what happens to the vegetation and herbivore production over this second 30-year period could be helpful.

Thus far we have considered an idealized setup in which the productive capacity of the vegetation remains constant from year to year. In realistic situations the amount of vegetation produced will fluctuate annually, dependent on environmental conditions. Hence copy the model into a new worksheet and set up a new column headed "ENVIR." Use the function =(**RAND**()+0.5)^0.5 to allow the productive capacity of the vegetation to vary randomly from 30% below to 22.5% above its mean potential. This will require replacing the fixed value for K_v in the vegetation production column with the product of K_v and the environmental state. Investigate how much difference this makes to the yield in terms of herbivore production over both the initial 30-year period and the next 30-year period. Due to the random sequence of environmental states, you will need to repeat the trials several times, reentering the same herbivore stocking density each time, in order to work out the potential consequences.

Appendix 6.1   Formulating an age-structured population model incorporating density dependence

Copy the age-structured model that was outlined in Appendix 4.1, projecting constant exponential growth, into a new worksheet in the same spreadsheet folder. Several structural modifications will need to be made to this model to accommodate the functional dependence of the survival rates on the abundance or density level.

First, assume that the age-specific survival rates are linearly dependent on the population size, that is, the survival function has the form $S = a_0 - bN$. The intercept a_0 represents the maximum survival rate when the population size is close to zero. The slope coefficient b represents how steeply the survival rate declines with increasing density. Hence you now require two rows of survival parameters, one row containing the values for a_0, and the other the values for b. Set the maximum survival rate to 1.0 for all age classes, assuming that under ideal conditions at low density there is no mortality, except at the end of the life span.

Thought needs to be applied to appropriate values for the slope coefficients. In the exponential growth model, you had assumed annual survival rates of 0.5 and 0.9 for the juvenile and adult classes, respectively. Accepting that this is a reasonable combination, a reduction in juvenile survival from 1.0 to 0.5 is associated with a reduction in adult survival from 1.0 to 0.9, that is, the decline in juvenile survival is five times as steep as that in adult survival. Accordingly, assign a value of 0.001 to the slope coefficient for juveniles and 0.0002 to that for the remaining age classes. This means that juvenile survival will become zero at a population size of 1,000, at which stage the survival rate for the older age classes will be 0.2. However, the population growth rate will become zero before the survival rates become this low. The population size at this stage can be found when the model is run. It emerges from the effect of rising density on the overall set of vital rates. You will assume that the fecundity rate remains constant, unaffected by density.

Second, assume that the population is censused some months after the young are born, so that you can observe how the relative proportions of juveniles and adults change as the growth trends shift from increasing to stable. Little change would be evident in the juvenile proportion at the time of the birth pulse because density only affects the survival rate of the offspring after birth. Accordingly, the summed reproductive contribution to age class 0, obtained from the product of the number of females and fecundity for each adult class, needs to be multiplied additionally by the effective survival rate of the offspring. This survival term is obtained by inserting the appropriate parameter values for a_0 and b into the linear survival function. The juvenile survival rate effectively represents survival from conception through to the time after birth when the census is done, including fertility failures.

The third change needed is to derive the reproductive contribution from adult females in the same year, rather than from females in the previous year. This avoids having to incorporate additional density-dependent functions for maternal survival into the reproductive contribution. Hence you are assuming that an offspring cannot survive independently of its mother. With calculations done this way, fecundity for age-class one should be set to zero, that is, females give birth to their first offspring at age two.

Last, incorporate density dependence into the formulae used to calculate survival from one year to the next for the remaining age classes. In order to be consistent with the way in which contributions to the juvenile class are calculated, the parameter values refer to the age class entered, that is, the parameter values above the column for age class n_1 are for survival from age class n_0 into class n_1. Having entered these formulae, ensuring parameter references are anchored in the appropriate cells, copy them down over the time range to 30 years. You should find that the population growth rate declines from an initial value of around 50% per year toward zero. To reduce fluctuations in the age structure in the first few years, set the starting number of females in each age class as follows: N0: 6; N1: 4; N2: 3; N3: 2; N4: 1; N5: 1.

A plot of the change in total population size over time should show a sigmoid curve resembling that generated by the logistic equation. To confirm that the population growth matches that generated by the logistic equation, convert the graph plotting the relative population growth rate against time to one showing the change in the relative growth rate against the population size. Remember that this requires an XY (scatter) plot, but with points connected to depict the trend. You should see a linear decline in the population growth rate toward the zero growth level, although small jiggles may be generated by the change in age structure as the population stabilizes.

Add two further columns showing how the adult and juvenile survival rates change toward the zero growth level. For the suggested parameter values, zero growth should be associated with a juvenile survival rate of 0.38 and annual adult survival of 0.88. Generate a graph showing how these class-specific rates change with increasing population size.

Finally add two more columns to reveal how the juvenile and adult proportions in the population change with the shift in the population growth rate. The juvenile proportion is represented by the number for the column N0, divided by the population total, while adults encompass classes N2–N5. Plot how the population composition changes over time as the population grows. In general, an increasing population contains a high proportion of juveniles, and a stable population a high proportion of adults. Experiment by changing the slope coefficients for either juvenile or adult survival, or for one of the adult classes, and observe how this changes the zero growth level as well as the population structure at this level.

Explore further by making the adult survival rate a convex function of the population size through incorporating a power coefficient $z > 1$ in the density-dependent equation $a = a_0 - bN^z$. Start by copying the current model into a new worksheet. An additional line at the top to contain the parameter setting for z is needed, and the value of b should be reduced to accommodate the inflation of the effective value of N produced by the power coefficient. Set $z = 3$ and make $b = 0.0000000005$ (i.e., nine 0s before the 5), for the adult classes. Retain the linear function unchanged for the juvenile class. You should find that the density dependence in the overall population growth rate is also now convex. However, the combination of vital rates associated with zero growth changes very little.

Complete your exploration by adding an additional adult class N6, representing senescent animals with a reduced survival rate. Make the maximum survival rate for this class equal to 0.8 of that for prime-aged adults, and the slope of the decline in survival rate with density twice as steep as that for prime adults. Also set the fecundity of this class to be 0.8 times that of prime adult females. Allow animals to persist in the senescent stage indefinitely by adding animals surviving in it from the previous year to animals moving into this class from the age class below. Do not forget to add the contribution of the senescent class to the juvenile class as well as to the total population and adult segment. You should find that the senescent stage does not have much influence on the zero growth level or other features because few animals survive in it.

Appendix 6.2 Incorporating environmental variability into the age-structured model

The effect of environmental variability on the class-specific survival rates can be incorporated simply by making the slope coefficient b a function of the state of the environment. This is equivalent to making the carrying capacity K vary with environmental conditions as was done for the descriptive population models in Chapter 3.

Copy your previous age-structured model into a new worksheet, and add an extra column headed ENVIR. Use the function =**RAND**()+0.5 to make the value of the random number vary uniformly between 50% below and 50% above the mean. Then alter the formulae for calculating survival for each age class by multiplying the cell reference to the value of b by the state of the environment in that year, and copy the new formula down.

However, the linear survival function may generate values exceeding 1.0 or falling below zero in some years, which is nonsensical. You will need to add an IF ... THEN condition to reset values to these limits if they are exceeded. In practice, because of the different slope coefficients, the juvenile survival rate is unlikely to exceed 1.0, while the adult survival rate is unlikely to drop below 0. Even if the juvenile survival were to exceed 1.0 in the first few years while the population is small, this could be interpreted as simply an effect of an elevation in fecundity.

Hence in the column calculating juvenile recruitment into the first age class, enter the following formula, assuming that the state of the environment is represented in column B:

$$= \textbf{IF}((\textbf{C\$17} - \textbf{C\$18*B140*I140}) < \textbf{0,0,(D141*D\$16}$$

$$\textbf{+E141*E\$16} + \textbf{F141*F\$16} + \textbf{G141*G\$16} + \textbf{H141*H\$16)}$$

$$\textbf{*(C\$17} - \textbf{C\$18*B140*I140))}$$

Interpret it as follows: if the value for juvenile survival, calculated in the first term in brackets, is less than zero, it gets assigned the value zero, else its value is calculated from the reproductive contribution from all adult classes multiplied by the calculated survival rate for juveniles. To perform the same adjustment for the next age class, enter the following formula:

$$= \mathbf{IF}((\mathbf{D\$17} - \mathbf{D\$18}^{*}(\mathbf{\$B21}^{\wedge}\mathbf{0.5})^{*}\mathbf{\$I21}) > 1, \mathbf{C21}, \mathbf{C21}$$

$$^{*}(\mathbf{D\$17} - \mathbf{D\$18}^{*}(\mathbf{\$B21}^{\wedge}\mathbf{0.5})^{*}\mathbf{\$I21}))$$

with suitable adjustment in the column reference for each successive adult class. According to this calculation, if the calculated survival rate is greater than 1, the number surviving is equal to the number in the previous class a year earlier (i.e., all survive), else the number in the previous class is multiplied by the calculated survival rate. One further modification has been introduced. The factor for the state of the environment from column B has been transformed using a square-root power, to reduce its amplitude of variation. This was done because it seems reasonable to assume that the adult survival will be less affected by environmental variability than the survival rate of juveniles.

Copy these formulae down, and note how the graphical output changes. It will now be more meaningful to plot the annual variability in adult and juvenile survival rates against time, rather than against population density. Also, the plot of the relative population growth against the population size should be changed to a scatter of symbols rather than connecting lines.

Make similar changes to the model assuming adult survival to be a non linear function of the population size. Observe how the output from this model differs from that of the model assuming linear density dependence in adult survival. Finally, this model could be modified to replicate the population structure, functional relations, and parameter values incorporated into the True BASIC program for a kudu population on the CD.

Appendix 7.1   Harvesting model based on the logistic equation in a constant environment

Provide a name for your worksheet at the top, then set aside cells to contain the values of the basic model parameters r_0 and K. Choose appropriate values for the population that you have in mind. Then enter the following column headings: "YEAR," "PRE-HARVEST POP," "POST-HARVEST POP," "RELGROW," "RECRUITM," and "HARVEST." Enter a starting value for the preharvest population in year 0 equal to the carrying capacity K. To calculate the postharvest population, subtract the harvest quota, which at this stage is undefined. Enter the logistic formula to calculate the *relative* growth rate as a linear function of the *postharvest* population, and then the annual recruitment as the product of the relative growth rate and postharvest population size. Start with an initial harvest quota of zero, and copy all of the formulae down over a 50-year time range. Both the pre- and postharvest population should remain fixed at the level set by the value of K.

Then enter an annual removal quota somewhat less than the theoretical MSY (calculated as $r_0 K/4$). Observe what happens to the population size over time, both pre- and postharvest. The population should decrease to a level somewhat below K, and then remain at this level indefinitely. To see this, produce a graph showing the change in abundance over time, for both the pre- and postharvest populations. Raise the annual harvest toward the MSY level, and then exceed this. All harvest quotas below MSY should be sustainable. Once the annual offtake exceeds the MSY, the population declines progressively over time, and will ultimately be extirpated, unless removals are suspended. Whether the initial downward trend continues indefinitely may not be clear over the 50-year period. To see beyond it, first reduce the postharvest population to the level giving the MSY, that is, half of the potential carrying capacity, by setting a harvest quota equal to $K/2$ in the first line. In the next line, set the harvest quota to be maintained each year from then on. If this annual quota is below MSY, the population will increase from this level. If MSY is exceeded, the population will decline progressively from the initial level. Investigate further what happens when the initial abundance level is set at or below $K/2$. The population should either grow toward a higher sustained level or decline toward extinction, depending on whether the annual removal you applied could be supported by this population size.

Appendix 7.2   Harvesting model for a goose population accommodating environmental variability

Start by copying the simple harvesting model outlined in Appendix 7.1 into a new folder. One additional column needs to be inserted, to contain a value for the annual rainfall. From the linear relationship between the population growth rate and population density for magpie geese established by Bayliss (1989; see Fig. 7.3a), r_0 is estimated to be 0.78 per year and K 250 geese km^{-2}. The realized growth rate would be also linearly dependent on the relative variation in rainfall around the mean, with a slope coefficient close to 1.0 (Fig. 7.3b). Furthermore, these two effects seem to contribute independently. This means that if the rainfall for the year was 20% above the mean, the relative growth rate was 0.2 higher than expected for the prevailing density. The highest observed annual growth for the goose population was 0.55 per year, and it is questionable whether the actual growth rate could much exceed this if the population fell lower than the minimum observed level. In practice, whether or not the annual increase is truncated at this maximum (e.g., by a conditional "IF" statement) makes little difference, since the population rarely declines to such low levels.

In setting up a discrete-time model, it is important to be aware of the order in which different events occur, in particular, the timing of removals in relation to the timing of reproduction. Hunting generally occurs in autumn, after the young geese have fledged, and the birds that survive the hunt contribute toward replacing those killed, through their reproductive output. Compensation occurs because the density of pairs on the breeding grounds has been reduced. This is why the annual increase is generated by the postharvest population.

For the regression relationship depicted in Fig. 7.3b, rainfall is expressed as a relative variation around a mean of zero, with a CV of 20% associated with mean annual rainfall exceeding 1,000 mm. To represent this in the worksheet, in the rainfall column, type the following expression: =((**RAND**()+0.5)^B$12)−1, with the power coefficient referenced in cell **B$12** having the value 0.65 to reproduce the above CV. Copy this formula (and all of the others) over a 100-year period. Alter the formula for the relative growth rate to incorporate the rainfall influence, by multiplying the slope coefficient for the rainfall influence on population growth (which happens to be approximately 1.0) by the adjusted random number appearing in the rainfall column. It is helpful to specify the slope parameter of 1.0 in one of the cells at the top of the spreadsheet, so that you could switch the rainfall influence off by changing its value to zero in order to observe the dynamics in a constant environment (provided of course that the formula refers to this cell). Copy this changed formula down. At this stage leave the population unharvested. Graph how the simulated goose population fluctuates between about 40% below and above its zero growth level of $250 \, km^{-2}$, corresponding closely with the observed range of variation in abundance. To investigate the consequences of more widely variable rainfall, raise the power coefficient for the rainfall measure from 0.65 to 1.0, thereby elevating the annual CV in rainfall to around 0.3. The amplitude of the fluctuation in the goose population is consequently much greater, especially below the mean.

To add harvesting without losing this basic pattern, copy the model into a new worksheet. Start by switching the rainfall effect off by making its slope coefficient zero, so that you can establish the harvest quotas that could theoretically be sustained in an idealized constant environment. For the specified parameter values, MSY is 39% of 125, that is, an annual harvest of just under 50 geese km^{-2}.

Then restore the rainfall influence, and establish what reduction in the annual harvest is necessary to ensure that the population can be sustained. A 100-year time horizon is rather remote, so reduce the period spanned in the plot of the population trend to 50 years. You will have to guess at the reduced quota, as there is no ready formula to guide you. Moreover, for each assigned quota, you will need to undertake repeated trials because of the random variability in the rainfall. To get a new random number sequence, press the function key F9, or make any entry in a cell of the worksheet. You will also need to decide what outcome you interpret as exceeding the sustainable level. In some trials population may have dropped well below the threshold density of $K/2$ after 50 years, indicating that it is certainly headed toward extirpation, provided the quota remains unchanged. To observe the ultimate trend, start with the population slightly above $K/2$, otherwise all that will be depicted over the first several years is the trend from carrying capacity toward the lowered density associated with the harvest.

You will find that the population persists over some 50-year periods, and crashes out in others, despite the same fixed quota, depending on the specific pattern of the rainfall. Hence for each assigned quota, undertake 100 trials, and work out the proportion of 50-year periods during which the population crashed. You will have to weigh the risk of having a crash in the next 50 years against the benefit of securing a higher annual harvest over this period if the population does not crash. Having established the risk–benefit relationship for a rainfall CV of around 0.2, increase the rainfall CV to 0.3 and ascertain what further reduction in the annual harvest is needed to have a sufficiently high confidence that the population will be sustained for 50 years.

To assess alternatively the benefit of assigning the quota as a proportion of the population, rather than as a fixed number, copy the previous model into a new worksheet. Since the permissible harvest must be assigned in advance, the proportion has to be based on the postharvest population in the year preceding that in which the harvest is undertaken. Specify the formula to calculate this quota in the harvest column. Compare both the size of the mean annual harvest and the risk of a population crash for different quotas with that obtained from a fixed quota, for different rainfall regimes.

Appendix 7.3   Simple dynamic pool model for fish harvesting

Set up the spreadsheet with a name and space for parameters and graphs at the top. Then set up column headings for "TIME" in years, "ENVIR" state, and four recruitment stages for the fish population. Label these stages "EGGS," "FRY," "SMALL," and "MATURE." It is helpful to distinguish the production of eggs, based on the fecundity parameter, from the subsequent survival of the fry after hatching. For fecundity, enter a value of 1,000 above the column for mature fish. This means that each mature fish produces 2,000 eggs, half of which give rise to female fry. Obviously only a small fraction of these fry survive very long after hatching, otherwise the population increase would be enormous. Enter the following values for annual survival rates above the respective columns: eggs (to become fry), 0.03; fry (to become small fish), 0.2; small fish (to become mature fish), 0.6; mature fish (to remain as mature fish), 0.6. These are the maximum rates under ideal conditions. They yield a maximum proportional growth rate of 1.22 per year, that is, the population can more than double between one year and the next, under ideal conditions at low density.

Two processes restrict this growth potential: (i) a density feedback through cannibalism by adults on the fry after hatching and (ii) variability in environmental conditions affecting the survival of the fry after hatching, presumably related to water temperature. Assume that the density feedback on fry survival follows the form of the Beverton–Holt equation, that is, $S = S_{max}/(1 + aN_t)$, where S is the effective survival rate, S_{max} is the maximum survival rate, N_t is the effective abundance level, specifically of adults, and a is the coefficient controlling how steeply the survival rate declines with increasing density N_t. In practice, it is more convenient to replace a by its inverse, which I will label k (i.e., $a = 1/k$), because increasing values of k raise the abundance level at which the population growth rate reaches zero.

Insert arbitrary starting numbers for these population segments at time zero. Then head three further columns "STOCKGROW," "STOCKRECR," "DDF," and "HARVEST." Insert a value for the parameter k above the heading "DDF," say 500. Then enter the formula to calculate the value of the density-dependent factor as $1 + N_t/k$, referring to the number of mature fish for the value of N_t. The number of mature fish also represents the size of the harvestable stock, assuming that the mesh size of the fishing nets is adjusted accordingly. Hence the annual growth in the stock is calculated as $(N_{t+1}/N_t) - 1$, as a proportion. It will not have a value at this stage because the size of N_{t+1} is not yet defined. The recruitment to the stock is most simply calculated as the difference between the stock size next year and the stock size this year. Initially set the annual harvest to be zero.

The number of mature fish surviving from one year to the next is the sum of the *postharvest* number of mature fish left from the previous year, times their survival rate, plus the number of small fish from one stage earlier surviving into the adult stage. Hence the harvest total should be subtracted from the number of fish in the mature stage in the previous year in this calculation. For other stage classes, the number of fish surviving is simply the number in the previous year times the survival rate between these stages. The number of eggs produced depends on the number of mature fish that survived from the previous year, and escaped being caught. Hence calculate it by multiplying the number of mature fish in the SAME year by the fecundity rate for this stage. Similarly, since the fry emerge in the same year that the eggs were produced, multiply the number of eggs by the survival rate for fry produced, also in the same year. However, since this is where the density feedback takes effect, the basic survival rate for this stage needs to be *divided* by the DDF value for this year.

You need to allow additionally for the effect of the environment on the survival of the fry. Initially set the state of the environment to remain constant at a value of 0.5. Then adjust the proportion of fish surviving from eggs into the fry stage by multiplying the previously entered formula by the number appearing in the "ENVIR" column for the same year. Having entered the relevant formula in each of the columns for time 1, copy these formulae down over a 50-year range, remembering to anchor the cell references to the basic parameters values to the cells containing them.

To view the intrinsic dynamics of the population, make the initial numbers in each stage very small so that the population can grow. Generate two graphs, one showing the change in the stock over time, the other showing the change in the proportional growth in the stock, together with the annual recruitment to the stock, against the stock size. Obviously one of these plots will need to be shifted to the secondary Y-axis to be visible. You should find that the stock size increases sigmoidally toward some upper limit under 2000, even though the density feedback affects only one stage. Note also that the relative growth rate of the stock at very low abundance is only about half of its maximum potential because the median state of the environment is below the ideal. Observe that the decline in relative growth rate with increasing density deviates from the linear relationship that you would expect from the logistic model, with a corresponding shift in the location of the peak in the recruitment curve relative to the maximum stock level. From the recruitment curve, you can obtain estimates of the MSY and the stocking level that gives this yield. Of course, these are ideal values in a constant environment.

Now set the starting numbers in each stage at time zero to be equal to the numbers at the bottom of the table after the population growth has stabilized at zero. This avoids any initial oscillations due to adjustments in the age structure to this steady state. Set the annual harvest quota to a value slightly below $rK/4$ (getting these values from the graph). Establish that this quota is sustainable under these idealized conditions. To avoid negative numbers produced by harvesting more than the actual number of fish in the mature class, you will need to adjust the formula for the harvest and for calculating the number of mature fish a year later. This requires entering the following condition: IF the number of mature fish exceeds the harvest for that year, it is calculated as the sum of the contributions from surviving mature fish and small fish growing into this stage; OTHERWISE it is calculated solely from the number of small fish, assuming all mature fish were removed. Similarly, if the size of the assigned harvest exceeds the size of the stock, the harvest actually removed becomes the total size of the stock in that year. It will help to have a third graph, showing how the annual harvest changes over time. Also calculate the mean annual yield over the 50-year period in a cell adjoining the HARVEST column. You should find that there is actually no limit to the quota that can be set, provided smaller fish are protected from harvesting. Even if the total stock is removed each year, it gets replaced by fish growing into this stage from the smaller size class, and the way the model is set up allows these fish to reproduce before they get removed by catches timed to occur after the breeding season. However, at some point the mean annual yield does not increase with further increases in the quota because the entire stock is being removed each year.

Then incorporate environmental variability by entering the formula =**RAND**() in the "ENVIR" column, and copying it down. The environmental state thus varies uniformly between one, representing ideal conditions, and zero, with a mean value of 0.5. You should find that at some point the mean annual harvest obtained decreases with further increases in the harvest quota. This occurs when the stock abundance gets depressed so much that it cannot recover from a bad year when recruitment was low. Note further that this overharvesting is not apparent in the first few years because harvesting started with the population at the high abundance maintained in the unharvested state. To establish whether a particular harvest is sustainable early on, the population should initially be reduced to a level around one half of its original abundance, that is, toward the level giving the theoretical maximum production. Observe how the yields change when you undertake repeated trials for the same quota because of the random sequence of environmental conditions that happens to occur. Hence you cannot be sure whether a particular harvest level is sustainable or not over a 50-year period.

Next investigate the consequences of fixing the harvest effort, thereby removing a constant proportion of the stock each year. Copy the model across into a new worksheet, remembering to adjust the source data for the embedded graphs accordingly. Assign the annual harvest as a proportion of the stock, and establish the maximum annual yield that can now be sustained. While there should be much greater security against a population crash for the same annual yield, the harvest obtained will fluctuate quite widely between years. Last, explore the consequences of a fixed escapement strategy. This entails setting the quota based on the size of the stock above some baseline level. This policy is guaranteed to be safe against overexploitation and should give the highest average yield over a long sequence of years. However, the annual harvests themselves become even more variable. You will now need a conditional "IF" statement in the harvest column to ensure that the annual quota does not become a negative number, with strange consequences. Establish the escapement level that give the maximum average yield.

Appendix 7.4 Dynamic pool model for a long-lived fish with several recruitment stages

Copy the simple fish harvesting model described in Appendix 7.3 into a new folder. Insert two columns between the small fish and mature fish stages, to represent medium fish and large fish, respectively. Retain a fecundity of 1,000 for mature fish, and allow large fish to produce eggs with a reduced fecundity of 200. Set the survival rate for mature fish and large fish to be 0.8; for medium fish growing into the large stage, 0.65; for small fish growing into the medium stage, 0.5; for fry becoming small fish, 0.3; and for the maximum survival rate of fry hatching from eggs, 0.05. Reduce the sensitivity of fry survival to environmental variability by modifying the random number using a power coefficient less than one, for example, make it $=RAND()^0.5$. Copy down this formulae and those used to calculate the stage transitions, anchored to the parameter settings in the appropriate cells. The maximum population growth rate for these parameter values should be 0.66 per year.

The fish now take 4 years to become fully mature. Two stage classes now contribute to the reproductive contribution in the form of eggs, as well as to the density feedback affecting fry survival: large adults and fully mature adults. Being smaller, the former have a reduced impact on fry survival. Hence insert a new column to calculate the effective weighted abundance of this combined reproductive class, with the contribution of the large adult stage down-weighted by a factor of 0.67. Alter the density-dependent factor affecting fry survival to refer to this new column. Allow also for a similar density influence on the survival of fry into the small fish stage.

The harvestable stock is now constituted by two adult stages. Assume that the catch is partitioned between them according to their relative proportions in the population. Due to the difference in size between these stages, the effective yield needs to be transformed to become the combined biomass of the catch, down-weighting the large stage by a factor of 0.67. Add a new column to do this. Add a further column to calculate the relative size of the fish caught, calculated as the ratio between the yield and the harvest assigned as a total number of fish.

Having set up the basic model, explore the consequences of different harvesting levels and policies. You should find that the main effect of over-harvesting is to reduce the yield by shifting the fish caught toward the smallest stage in the stock pool. To view this effect, add an additional series to the graph showing how the harvest changes over time. This pattern can also be revealed by the mean size of the fish caught plotted on the secondary Y-axis in the same graph.

Appendix 8.1 Demographic stochasticity in a spreadsheet

This model takes into account whether each individual in the population either survives or fails to survive, based on a probability level. The reproductive contribution is considered in the same way, first, from the probability that the offspring will be female, and, second, from the probability that the offspring will survive to the time when the population is enumerated. The model will be kept very simple, and will consider only very small populations because spreadsheets cannot easily handle the contingencies that arise with larger populations or extended time periods.

The basic model assesses the likely population size one time step later, based on the initial size and particular probability settings for the vital rates. Set up the model in a new worksheet with a heading, and space under it for cells containing the fixed parameter values. Just two numbers are needed; initially make the annual survival probability 0.8 and the expected fecundity 0.25. The latter is derived from a 0.5 sex ratio followed by a 0.5 chance of the offspring surviving. These values were selected because they result in an expected population growth rate of zero, assuming that offspring becomes adult and reproduces one year after birth. Check this by incorporating these numbers into a matrix model with two stage classes.

Instead of just one column for the population size, you will need a column for each individual in the population. Start simply with just two. Hence label column A "TIME," and the next two columns "I1" and "I2," respectively. Head the fourth column "TOTAL." Under the heading "TIME," enter the words "Start," "Survive," "Reproduce," and "End" in the next four rows. Hence, having entered the starting numbers, you first calculate which individuals survive, then their reproductive contribution if they survive, and finally the population numbers at the end of the time step that result.

To decide whether each individual survives or dies, the assigned survival probability must be related to a randomly chosen number within the range 0 to 0.999 (rounded to three decimal places). The logical deduction is, IF the random number is less than the survival probability, the individual survives, ELSE it dies and is removed from the population. A similar procedure can be used to decide whether it contributes a surviving offspring, based on the effective fecundity.

To avoid cluttering the population model, generate these random numbers in a separate worksheet. To anticipate future needs, establish a set of cells spanning 20 rows and 20 columns, each containing the formula =**RAND**(). A set of numbers between 0 and 1, each different, should appear in these cells. If you press the **F9** key, or make any entry into a cell in the spreadsheet, the values of these numbers will change.

Return to the previous worksheet containing the population model. Begin with two individuals by entering the number "1" in the row labeled "Start" in the columns for each of these individuals. In the next row labeled "Survive," enter the following formula to determine whether individual 1 survives or not: =**IF(Sheet2!B12<Sheet1!$C$7,Sheet1!B11, Sheet1!B11*0)**.

The logic of this formula is, IF the value of the random number appearing in sheet 2, cell B12 is less than the value for the survival probability shown in cell C7, this individual survives, else it is eliminated by multiplying its starting number by zero. Adjust the cell references to those where you have entered the effective parameter values and starting numbers. Copy this formula across into the column for individual 2. Note whether the number 1 (meaning still present) or 0 (meaning gone from the population) appears in the row for "Survive." Press the **F9** key several times, and observe how these numbers switch between 1 and 0, depending on the changing values of the random numbers referenced in the second worksheet.

In the next row labeled "Reproduce," enter a formula of this form for each individual: =**IF((Sheet2!B13<Sheet4!$C$8),Sheet4! B12,0)**.

If the value of the random number is less than the effective fecundity, the offspring is both female and survives, otherwise no surviving offspring is contributed. The "End" population is calculated as the sum of the surviving adults and offspring, and the "TOTAL" is the sum of the numbers appearing across each row.

The interesting aspect of the outcome is how many individuals survive one time step later, depending on the starting number and the survival and fecundity settings, and in particular how frequently the population falls to zero. Press the **F9** key several times to observe how the end total changes.

Since the outcome is *stochastic*, that is, governed by chance as controlled by the particular set of random numbers, many trials are needed to establish the likelihood or frequency distribution of different end numbers. A hundred trials would give you an answer to the nearest 2–3%. A thousand trials would be needed to obtain a more precise estimate of the frequency distribution of outcomes.

These trials could be performed by pressing the **F9** key 100 or more times, writing down the outcome each time on paper or in a cell in the spreadsheet. But this can be rather tedious. Spreadsheets allow you to automate the sequence of key presses needed to get the outcome, by means of a *macro*.

The macro is set up as follows. First choose where you want the numbers representing the outcomes to appear in the worksheet. I will assume that they appear in a column, headed "TRIAL," starting in cell B20, and that the end total for the population size in each trial is in cell D14. The macro records the key presses that you use to copy the values of the end number from successive trials in the column headed "TRIAL," by repeatedly pressing a chosen "shortcut key."

Start by going to the "Tools" menu, select Tools/Options/Calculations, then the option "Manual Calculation." Next select "Macro" from the "Tools" menu, then "Record" mode, then assign the shortcut key (say CTRL z). Thereafter carry out the following steps to create the macro:

- Press **F9** (to change the set of random numbers)
- Select cell D14, then open Edit/Copy (to copy the end total)
- Select cell B20 (to identify the column heading where these numbers will be pasted)
- Open Edit/Find, choose Options, then Search "By Columns." Click "Find Next," then close (so that the numbers appear downward in the column)
- Open Edit/Paste Special/Paste Values, then click "OK"
- Finally, select Tools/Macro/Stop Recording.

Now when you press the shortcut key 200 times, 200 numbers representing the end population total will appear in the column headed "TRIALS."

Appendix 9.1   Basic spreadsheet model for metapopulation dynamics

Give your worksheet a heading as usual, then set aside cells to contain the model parameters. These parameters are the expected extinction rate E and expected colonization rate C. Assign a starting value of 0.1 for extinction, and 1 for colonization. Further down choose a block of cells 5 columns wide and 5 rows deep to represent a set of 25 habitat patches. Habitat patches that are occupied by a population will be represented by "1" and those that are empty by "0." Start by making all of the patches occupied.

You want to assess how patch occupancy changes over time. This requires replicating the mosaic of 25 patches over a sufficient number of time steps. First establish a column headed "TIME," and enter the value 0 in this column alongside the starting patch mosaic. In another column headed "INCIDENCE" adjoining the patch mosaic, calculate the proportion of patches occupied by summing across the 25 cells and then dividing by the cell number, that is, 25. The starting proportion should be 1.0, if you entered the number 1 in each cell. Below this initial patch mosaic, assign another 5 × 5 block of cells to represent the state of the patch mosaic in the next time step. Each cell requires a formula to establish whether or not it supports a population, based on state in the previous time step and the relative chances of colonization and extinction. Due to the stochastic nature of the state transitions between occupied and vacant, and vice versa, a random number is needed to determine the actual outcome for each cell. To avoid cluttering the worksheet, set up these numbers in a separate worksheet named "Random" in the same folder. Fill 200 rows and 20 columns with the formula =**RAND**(), in order to anticipate future needs. Return to the first worksheet, and in the first cell of the habitat patch mosaic for time 1, enter the following formula:

$$= \textbf{IF(A10} = \textbf{0, IF(Random!A10} < \textbf{\$C\$7*E15, 1, 0),}$$

$$\textbf{IF(Random!F10} < \textbf{\$E\$7, 0, 1))}$$

This is interpreted as follows. If the corresponding cell at the previous time step, referenced as A10, is vacant (equal to 0), it get colonized, provided the value of the corresponding random number in the other sheet is less than the colonization probability referenced in cell \$C\$7, multiplied by the proportion of patches occupied referenced in cell E15, otherwise this cell remains zero. If it was formerly occupied (i.e., the condition that it has a value of 0 is false), it becomes vacant (gets a value of 0), provided the value of a second random number is less that the extinction probability referenced in cell \$E\$7, otherwise it remains occupied (retains a value of 1). Suitably anchored as shown above, this formula can be copied across the remaining cells in the patch mosaic. You should observe that 2–3 of the patches that previously had 1s now get 0s. Calculate a value for the patch incidence at this time.

Next you need to replicate this procedure for determining the patch occupancy over additional time steps; 20 steps should be sufficient to move far enough from the initial patch setting. If you assigned the parameter values suggested above, you should find that two or three of the 25 cells contain zeros after 20 iterations. Obviously the exact expected proportion of vacancies (calculated as $1 - [0.1/1]$ or 0.90) cannot arise because the tally must equal a whole number. To confirm that the results match expectations, first set the initial patch incidence to match the expected proportion, then calculate the average patch occupancy over the 20 following time steps. Undertake repeated trials, using the F9 key, and observe how much this number fluctuates around the expected mean of 0.90. It is also helpful to graph the change in patch incidence over time.

Reduce the value for the colonization rate C, and note how the metapopulation tends toward a different proportion of occupied patches. Even when you make C zero, many patches may still be occupied at the end of the iteration period, when extinction rate is low. To observe the final outcome, start the next trial with the patch proportions at the end of the previous one, and repeat this procedure until the patch incidence remains constant.

For visualization, it is helpful to shade the occupied cells using some color. For this purpose, establish a 5×5 patch mosaic somewhere near the top of the worksheet, and reflect the values (1 or 0) of the cells at the end of the iteration period into it (i.e., simply make the cell contents equal to those of the other set of cells). Select these 25 cells with the mouse, then go to the Format menu at the top of the spreadsheet, and set up the following conditional formatting from the table that appears on the screen: if the cell value is equal to 1, give it an appropriate color, say bright green. Then each time you press F9 to undertake a repeated trial, different cells in this array will show up as bright green, indicating that they support a population at the end of the 20 iterations.

The model is now set up to allow an exploration of the consequences of different values of C and E, and also of reducing the number of patches in the mosaic. If you make C equal to E, you should find that extinction of the whole metapopulation occurs occasionally when you press the F9 key a sufficient number of times, even with 25 patches. Furthermore, stochastic variation in the number of occupied patches sometimes lowers the effective colonization rate (i.e., the potential rate reduced by the proportion of patches occupied by populations) sufficiently so as to generate a form of "extinction vortex," leading rapidly to the disappearance of the entire metapopulation. This effect becomes amplified when you reduce the number of patches in the mosaic, because each patch extirpation then makes a bigger proportional difference to the colonization rate.

Appendix 9.2   Model representing correlated migration and extinction

To introduce correlated migration, just a small change to the model set up in Appendix 9.1 is needed. The cell reference to the random number determining colonization, underlined below, must be identical for all of the patches in each iteration over the 20-step period.

$$= \textbf{IF(A10} = \textbf{0, IF(Random!}\underline{\textbf{A10}}< \textbf{\$C\$7}^*\textbf{E15, 1, 0)},$$

$$\textbf{IF(Random!F10} < \textbf{\$E\$7, 0, 1))}$$

Copy the model previously formulated into a new worksheet, choosing a reduced mosaic of 10 habitat patches. Alter the formula in each cell of the patch mosaic, then copy this change across the successive iterations. Make the colonization rate equal to twice the extinction rate, and establish the probability that the metapopulation will still be in existence after 20 time steps. Compare this probability with that obtained using the previous model assuming independent migrations, for the same parameter settings.

Next introduce correlated extinctions. Again start by copying the model worksheet from Appendix 9.1 into a new worksheet. You now need to determine the overall likelihood of patch extinctions, taking into account the independent probability of local extinction as well as the probability of local extinction from spatially correlated disturbances affecting all patches. Recall that the probability of both A and B occurring is the probability of A times the probability of B. Hence to assess the joint probability of events A and B occurring, without changing the basic likelihood of local extinction set at the top of the spreadsheet, you need to calculate the geometric average, which is the square root of the product. Hence the formula determining patch occupation in each time step becomes

$$= \textbf{IF(A10} = \textbf{0, IF(Random!A10} < \textbf{\$C\$7}^*\textbf{E15, 1, 0)},$$

$$\textbf{IF(}\underline{\textbf{(Random!\$B8}^*\textbf{Random!F10) 0.5}} < \textbf{\$E\$7, 0, 1))}$$

The change needed is underlined. The random number anchored in cell B8 determines whether or not the correlated disturbance affecting all local populations occurs in that time step.

Having made this change, compare the likelihood of metapopulation extinction with that for an equivalent set of habitat patches governed by identical, but independent, local patch extinction rates. I suggest setting $C = 0.4$ and $E = 0.2$ initially, so that there is a reasonable chance of each patch becoming extirpated over the 20-step iterations within each trial. You should find much more variability in the outcome between successive trials, depending on whether or not the condition for a disturbance is met and how pervasive its effect is across the cells. Even if the extinction rate is lowered relative to the colonization rate, periodic disturbances can still cause the metapopulation to fade out in some trials.

Appendix 9.3   Model representing variable patch size and spacing

Copy across the model with correlated disturbances, but not correlated migration, into a new worksheet, for a situation spanning a 5×5 mosaic of cells. Reduce the number of these cells that actually represent habitat patches as follows. Establish a block of four contiguous cells in the upper left corner to represent the core habitat region, by entering the number one ("occupied") within them at time zero. Then establish starting populations in three cells at a moderate distance from this block, and a further three cells at a greater separation from the core block. The remaining cells should all be blank. Thus, in total 10 cells represent habitat patches while 15 cells are intervening areas that are unsuitable habitat, although allowing migrants to move between the patches.

Next the extinction and colonization probabilities of the 10 patches need to be differentiated, dependent on their size and spacing. Assume that the large population spread over the core habitat patch has zero chance of becoming extirpated through demographic processes or normal environmental variability, but has a slim risk of being wiped out by some extreme disturbance. Hence alter the formula in these four cells at time 1 as follows:

$$= \text{IF}(\text{A10} = 0, \text{IF}(\text{Random!A10} < \$C\$7^*\text{E15}, 1, 0),$$

$$\text{IF}((\underline{\text{Random}\$\text{B8}})^{\wedge}0.5 < \$E\$7, 0, 1)).$$

Note that the contingency for extinction (underlined) now refers only to the random number determining whether or not a correlated disturbance occurs. Furthermore, the square root of this single random number must now be less than the basic extinction probability specified in cell E7. This means that if E is set at 0.1, local extinction will occur only if the random number is less than 0.01 (the square root of 0.1).

The formula in the three cells at intermediate separation from the core block needs a small adjustment, to allow a possibility that one of these patches could retain a population even after a disturbance sufficiently extreme to eliminate the core population:

$$= \mathbf{IF(A10 = 0, IF(Random!A10 < \$C\$7^*E15, 1, 0),}$$

$$\mathbf{IF((Random\$B8^*Random!F10^*\underline{1.2}^{\wedge}) \; 0.5\_< \$E\$7, 0, 1)).}$$

For the three remote cells, the chance of colonization needs to be reduced. Hence make the following adjustment, to halve the colonization rate:

$$= \mathbf{IF(A10 = 0, IF(Random!A10 < \underline{0.5}^*\$C\$7^*E15, 1, 0),}$$

$$\mathbf{IF((Random\$B8^*Random!F10^*\underline{1.2})^{\wedge}0.5 < \$E\$7, 0, 1)).}$$

Set $E = 0.1$ and $C = 0.5$, so that on average 80% of the patches should contain a population, for a situation where the patches are equal in size and spacing. Start with all patches occupied. Note how in most runs the patch occupancy remains above 0.8 because of the remote chance of extirpation of the core population. However, about once in 100 iterations a disturbance occurs that is sufficiently severe to eliminate the core population along with the peripheral populations. If you run a very large number of trials, you should find occasional instances in which one of the peripheral populations survives such an event and reestablishes the core population.

Appendix 9.4 Model representing source and sink populations

Copy across the worksheet representing variable patch size and spacing, outlined in Appendix 9.3. Assume that the block of four cells in the upper left represents the source habitat patch. Chance extinctions following catastrophic disturbances are no longer of concern, so eliminate the extinction contingency in the formula appearing in these cells. This can be done simply by switching the cell reference for extinction from the random number sheet to a cell containing a number greater than the extinct probability, say 1. Hence the formula becomes

$$= \mathbf{IF(A10 = 0, IF(Random!A10 < \$C\$7^*E15, 1, 0),}$$

$$\mathbf{IF(\$F\$7 < \$E\$7, 0, 1))}$$

where the cell F7 contains the number 1. Also eliminate the contingency for correlated disturbances in the peripheral six patches, so that chance extinctions once more occur independently among the peripheral sink patches. Hence the formula for these cells becomes identical to that given in Appendix 9.1. Set the value for the colonization probability C to 1, meaning that some migration occurs in every time step, at least from the source habitat. Initially set the extinction rate E also equal to 1, so that all populations established in the sink patches become extirpated in the next time step. Undertake repeated trials with this situation, and observe that typically 1–2 sink patches retain populations at any one time.

Next reduce the extinction probability to 0.5, so that on average only half of the populations established in sink habitats go extinct in each time step. Observe how between zero and four of the sink habitats contain populations at any time. Note that because the effective colonization rate depends on the proportion of occupied patches, extinction and colonization rates are actually roughly equal, despite the difference in the parameter settings. Also explore how values of E between 0.5 and 1 alter the proportion of sink habitats containing populations.

Last, observe what happens when the source population is eliminated, retaining the above settings of C and E and all six sink patches initially occupied by populations. You should find that all of the sink populations fade rapidly into oblivion because migrants leaving them are inadequate to counterbalance an extinction rate equal to or greater than 0.5. Even if the extinction rate is lowered to 0.25, sink populations generally persist for less than half of any 20-step period.

Appendix 10.1   Modeling the outbreak dynamics of a transient infection: measles

Set up a worksheet with a heading and several lines left for parameter values. Establish five columns, headed WEEKS, SUSCEPTIBLE, INFECTED, RECOVERED, and TOTAL. The weekly time step corresponds with the duration of infectiousness of 7–10 days. Just two parameter values are required: the rate of transmission of the infection, symbolized as β, and the rate of recovery from the infection, symbolized as v. The rate of recovery is the inverse of the duration of infection, so set v to be 0.7 per week as an approximation. This means that 70% of those infected with measles have recovered with immunity 1 week later whereas 30% remain sick and capable of passing on the pathogen. The value of β depends on the effective size of the population influencing the frequency of contacts. Assume a total population of 1,000, representing a medium-sized city school. Initially set β equal to 0.001 per week, which gives an R_0 of 1.43. Start the population at time zero with 1 infected individual, 999 susceptible individuals, and 0 in the recovered category. The number of new infections is the product of the transmission rate, the number infected, and the number susceptible. Allowing for the fraction in the infected category that remain still infectious, the formula needed in the "Infected" column is

$$= B\$7^*B10^*C10 + C10^*(1 - D\$7)$$

where line 10 contains the starting numbers in week 0, cell B7 contains the fixed value of β, and cell D7 contains the fixed value of ν. The corresponding formula for the number of susceptibles is

$$= B10 - C10^*B10^*B\$7$$

while the number recovered is calculated by

$$= D10 + \$D\$7^*C10.$$

The total is simply the sum of susceptible, infected, and recovered, and should remain constant at 1,000. Copy the formulae over a 2-year, that is, 104-week period. Graph the changes in numbers in each of the three classes against time. You should observe that the number of infections rises rapidly to a peak, then fades equally rapidly to zero. With the above parameter settings, just over half of the population gets the disease over the course of the first year, and recovers with immunity. No infections recur during the second year, or in any subsequent year if you extend the period.

Experiment by varying the transmission rate β. If β is reduced to 0.0007 or less, the disease fails to spread. Make sure you understand why this is so. Increasing the value of β results in an earlier peak in infections and a higher proportion of the population becoming infected. If the value of β is too high, you will get negative numbers in the susceptibles column, suggesting that more than the entire population gets infected. This is an artifact of the discrete time step being too long for the rate of spread. Set a MAX(X, 0) condition in this column to ensure that values cannot fall below zero, where X represents the formula.

To allow for new individuals entering the population in the susceptible category, you will need a third parameter for the "entry" rate ε as a fraction of the total population. Copy the previous model into a new worksheet, and set ε to be initially 0.004 per week, amounting to about a 20% influx over the course of a year. Modify the formulae to add these new individuals in the "Susceptible" column , and subtract the same number from the "Recovered" column to keep the total population size constant. This assumes that all children have been infected with measles before they leave school. Copy these formulae over the 104 weeks, then extend the time range to span 10 years, that is, 520 weeks. Check that the number of infected individuals does not drop below one, otherwise there is no source of infections (although the model may allow fractions of individuals).

With β equal to 0.001 per week, you should observe recurrent outbreaks of infections occurring approximately every second year. If you start with a population completely susceptible, except for the initial infected individual, the first spike of infections will be much larger that those subsequently. To represent a steady-state situation, adjust the initial numbers in each column in week 0 to those prevailing when the infected proportion is lowest. Vary the values of β and ε, and note how raising either reduces the period between outbreaks.

To represent the synchronous entry of susceptible children into the school-going population at the start of each year, establish a third model in a new worksheet. Delete the weekly entry and loss of individuals from the formulae in the "Susceptible" and "Recovered" columns. Instead you need to represent a pulsed entry and exodus at the start of each year, that is, in weeks 53, 105, 157, and so on. Adjust the formulae in these specific weeks over the 10-year period to add new susceptibles and subtract an equal number recovered based on the entry rate converted to an annual scale. A combination giving an outbreak every second year is $\beta = 0.001$ and $\varepsilon = 0.3$. Adjust these values slightly and observe what difference the changes make to the pattern of the recurrent outbreaks.

Appendix 10.2   Modeling the dynamics of a sexually transmitted disease: HIV–AIDS

Set up new spreadsheet with the column headings YEARS, SUSCEPTIBLE, INFECTED, NEWINFECT, DIED, and TOTALPOP. The numbers in the columns should now be proportions rather than numbers. These proportions represent the population segment comprising sexually active adults aged between 15 and 49 years. Make the initial proportions 0.999 susceptible and 0.001 infected, which approximates the situation among mine workers in South Africa in 1987. Set the rate of transmission β of HIV to be 0.8 per year, and the duration of infection to be 7 years.

The proportion of new infections generated during the course of the year is given by the product $\beta*(\text{INFECTED})*(\text{SUSCEPTIBLE/TOTALPOP})$. The total population is equal to the sum of the proportions in the susceptible and infected categories, and should be equal to 1.0 initially. New infections are deducted from the proportion in the susceptible category in the following year, and added to the proportion in the infected category. The proportion dying may be calculated in two different ways. For the first worksheet, assume that the mean duration of infection before death of 7 years translates into a mortality loss amounting to one-seventh of the infected proportion per year. This means that how long each individual lives with the infection varies widely around the mean. The proportion dying is deducted from the proportion remaining in the infected category in the following year.

Copy these formulae down, and you should find that the infected proportion is approaching 35% after 12 years, that is, by 1999 assuming the starting year to be 1987. By this stage the annual mortality from the disease has reached around 5%, and the total population is declining. This means that, in order to calculate true proportions, the fractions appearing in the columns are no longer true proportions. Hence add two further columns, headed ANNMORT and INFECTPROP. Calculate the annual mortality as a proportion by dividing the fraction in the column headed DIED by the total population. Likewise, calculate the proportion infected by dividing the fraction in the INFECTED column by the total population. Graph the changes over time in these proportions as well as the population total as a fraction of the starting population. Note that the annual mortality rate reaches its peak after 17 years, by which stage almost the entire susceptible population has become infected, and the total population is dropping rapidly. Contrary to simple theory, infections continue to spread beyond the stage when the susceptible proportion drops below $1/R_0$, because deaths elevate the proportion susceptible within the segment remaining alive, while infected survivors continue to pass on the disease.

In reality, the population decline would be ameliorated by individuals born into the population entering the susceptible segment at 15 years of age. Hence the proportion in the susceptible column should be augmented by the product of the total population and the effective birth rate (assuming that infected individuals continue to produce children). A reasonable birth rate of 2% per year slows the spread of the disease slightly. Allowance can additionally be made for mortality from causes besides HIV. This entails multiplying the fractions calculated for the susceptible and infected categories by one minus the death rate. Explore the consequence of different combinations of birth and death rates. Elevating either slows the projected spread of infections.

Once the proportion infected becomes more than a few percentage, annual time steps become too coarse to represent the spread of infections and consequent mortality. Hence reduce the time step in the worksheet, say to one-tenth of a year, from this point onward. This means that the infection rate β as well as the birth and death rates need to be adjusted accordingly, for example, β of 0.8 per year becomes $(1.8^{0.1} - 1)$ per one-tenth of a year.

For a second version of this model in a new worksheet, assume that individuals die exactly 7 years after having been infected with HIV. This requires adjusting the calculations in the column headed DIED. The fraction dying is basically equal to the new infections that occurred 7 years earlier. A small adjustment is needed to allow for natural mortality between the time of becoming infected and when death from AIDS results. This entails multiplying the fraction of new infections by one minus the death rate, raised to the power seven (for the 7-year lag). With delayed deaths, infections spread more rapidly than in the previous model, approaching 30% after 10 years with the birth rate and death rate both held at zero. Furthermore, the annual death rate eventually rises to over 70% of the surviving, completely infected remnant of the original population. Allowing a birth rate of 4% per year counterbalanced by a death rate from non-AIDS causes of 2% per year reduces the peak death rate to around 30% per year, and continuing recruitment into the susceptible category restricts the maximum prevalence of infections to 94%. Once again, the projected proportions are slightly lower if a finer time step is used in the model.

Appendix 11.1   Modeling the white rhino–grassland interaction in a spreadsheet

This is the most complex spreadsheet model thus far, hence careful thought needs to be given to developing it incrementally. The first step is to formulate a model for the exponential growth of a white rhino population in a constant environment, to replicate the observed pattern up to 1972. Thereafter, density dependence in the vital rates, followed by the vegetation interaction, needs to be added to this model. Next, allowance needs to be made for the effects of annually variable rainfall. Finally, allow for density-dependent dispersal.

Head the opening worksheet "White rhino–grassland model," then enter the basic equations as text as a reminder. For vegetation dynamics, $dV/dt = rV(1 - V)$, where V is expressed as a proportion relative to some mean value (i.e., K is effectively 1). For the mortality loss incurred by white rhinos, $M = \exp(-a - b\ln(KH))$. For the natality function (births per adult female per year, i.e., the inverse of the calving interval), $N = -a + b(KH)$. For the intake function, $I = i_{max}(KH/(0.5+KH))$. For all of these equations, KH represents the maximum potential biomass attained by the white rhino population divided by the actual biomass level prevailing in a particular year, potentially modified by the state of the vegetation cover and the annual rainfall.

a. Exponential growth model

The spreadsheet should be set up with summary variables toward the left, where they are readily visible, and the detailed demographics of the white rhino population to the right. White rhino demography will be simplified by lumping all adults into one class, whatever their age, while separating recruitment stages among 10 year classes. These encompass both males and females, assuming a sex ratio of 0.5. The column headings needed are YEAR, WRBIOM, WRDENS, PGROW, ADULT, SA10, SA9, SA8, SA7, IM6, IM5, IM4, JUV3, JUV2, and JUV1. The parameters needed are simply the respective minimum mortality and maximum natality rates, as listed in Table 11.1. These rates are distinguished for the stages adult, subadult, immature, and juvenile, as indicated in the column headings. The class JUV1 represents newborn animals less than 1 year old. Body weights need to be assigned to these classes, in order to calculate the population biomass. These are, in order, 1,600, 1,520, 1,440, 1,280, 1,200, 1,120, 960, 800, 640, 480, and 240 kg. Start the population at time zero with 10 adults and 1–2 animals in each subsequent class. The number in each class 1 year later will be the number in the younger class multiplied by survival rate, which is one minus the mortality loss, between these classes. The adult class comprises adults persisting in this stage plus animals from class SA10 moving into the adult stage. The class JUV1 is derived from the adult and subadult stages multiplied by their respective natality rates, factoring in also the first-year mortality loss. Remember to include a sex ratio of 0.5 in this calculation because half of the adults are male. Calculate the effective population density (no. km^{-2}) as the summed population size divided by 100 (for numerical convenience), and the corresponding population biomass (kg km^{-2}) taking into account the body mass of each class. Calculate the proportional growth rate from the change in density between successive years. Copy the formulae down over a 30-year range. If you have entered the equations correctly, and used the parameter values for maximum population growth listed in Table 11.1, you should find that the population growth rate stabilizes at just under 9% per year. If not, look for and correct the errors you might have made.

b. Incorporating density dependence and vegetation interaction

Copy the above model into a new worksheet, then enter additional columns toward the left headed as follows: VEGPOP, VEGGROW, VEGPROD, KB,

WRCONS, XSUTIL, and WRINT. Above these column headings, additional parameter values need to be entered for the maximum annual growth rate of the vegetation (r_V, as a proportion), the amount of forage biomass potentially produced by the vegetation, (F_V, in $kg\,ha^{-1}$), the saturation biomass level for the white rhino population (K, in $kg\,km^{-2}$), the maximum annual food intake per white rhino (i_{max}, as a factor), and the regression coefficients a and b for the mortality and natality rates of each white rhino stage, as listed in Table 11.1. The minimum mortality and maximum natality settings also need to be retained. Since the stage-specific mortality losses vary depending on the population density and other factors, it is helpful to calculate these rates in additional columns: one each for adults, subadults, immatures, juveniles, and the JUV1 class. Add two further columns, one for adult natality and one for subadult natality. Calculate these rates using the formulae given in Table 11.1. Note that they require a value for KH, that is, the saturation biomass K divided by the actual white rhino biomass H in that year. To allow for lag effects, the effective KH values need to be averaged over the preceding 3- or 5-year period, which is best handled in two further columns. These rates also need to be constrained between their maximum potential value and zero, requiring conditional MAX or MIN restrictions in the formulae. Replace the constant mortality and natality rates that you assumed in the first worksheet by the variable rates calculated using the density-dependent formulae. Ignore the columns to the left representing the vegetation interaction initially, and copy the functions for the white rhino population dynamics down over a sufficient time span to reach the zero growth level set by the mortality functions. The biomass density attained may deviate slightly from the value for K that you set at the top of the worksheet because of the simplifications introduced. Plot both the change in population biomass over time, and the response of the population growth rate to changing population density. The former should show a sigmoid curve, and the latter should show an initial plateau followed by a ramp decline (see Fig. 11.4).

The starting value for VEGPOP, representing the grass cover relative to some saturation level, is 1. The vegetation growth is given by the logistic equation written at the top of the spreadsheet. Since the vegetation is already at saturation, initial growth should be zero. The vegetation production is obtained by multiplying the relative cover by the conversion F_V from cover to forage biomass. The annual food intake per unit white rhino is given by the equation for I, written at the top of the spreadsheet. The factor KH now needs to be multiplied additionally by the vegetation cover, implying that the carrying capacity for white rhinos is reduced when the vegetation cover is depressed.

Table 11A.1 Spreadsheet functions for the white rhino–grassland model

C	D	E	F	G	H	I
VEGPOP	VEGGROW	VEGPROD	XSUTIL	CONS	KB	WRBIOM
MAX(C20*(1 – F20) + D20,C$18)	D$16*C21*(1 – F21)*(1 – C21*(1 – F21))	E$16*C21*B21	MIN(IF(G21 >E21/2,(G21 – E21/2)/E21,0),F$18)	I141*J141/100	B21*C121* I$12/I21	(M21*M$17 + O21*O$17 + P21*P$17 + Q21*Q$17 + R21*PS17 + T21*T$17 + U21*U$17 + V21*V$17 + X21*X$17 + Y21*Y$17 + AA21*AA$17)/100

J	K	L	M	N	O	P
WRINT	WRDENS	PGROW	ADULT	ADM	SA10	SA9
MIN(J$12*H21/ (0.5 + H21),J$12)	SUM(M21:AA21)/ 100	K22/K21 – 1	(M20 + O20)*(1 – N20)	MAX(EXP(– M$11 – M$12 *LN($H21)),M$13)	(P20)*(1 – $S20)	(Q20)*(1 – $S20)

Q	R	S	T	U	V	W
SA8	SA7	SAM	IM6	IM5	IM4	IMM
(R20)*(1 – $S20)	(T20)*(1 – $S20)	MAX(EXP (–$P$11 – $P$12*LN($H21)), $P$13)	(U20)*(1 – $W20)	(V20)*(1 – $W20)	(X20)*(1 – $W20)	MAX(EXP(–$T$11 – $T$12*LN($H21)), $T$13)

X	Y	Z	AA	AB	AC	AD
J3	J2	J2M	J1	J1M	ADNAT	SANAT
(Y20)*(1 – $Z20)	(AA20)*(1 – $Z20)	MAX(EXP (–$X$11 – $X$12*LN($H21)), $X$13)	((M20*0.5*AC20)+ (O20 + P20 + Q20 + R20)*0.5*AD20)*(1 – AB20)	MAX(EXP(–$AA$11 – $AA$12*LN($H21)), $AA$13)	MIN((–M$14+ M$15*$H21), M$16)	MIN((–P$14 + P$15*$H21),P$16)

AE	AF
KB3	KB5

The consumption of vegetation by white rhinos is the food intake per capita I multiplied by the total white rhino biomass, and then divided by 100 to convert from kilograms per square kilometer (the units for herbivore biomass) to kilograms per hectare (the units for vegetation production). The excess utilization over the overgrazing threshold is calculated as the proportion by which the consumption exceeds the threshold, set at 0.5 of production. The formula needs to be set within a contingent IF so that its value remains at zero if the overgrazing threshold is not exceeded. Enclosing this whole expression within an additional MIN condition is also needed to ensure that the excess utilization does not exceed 0.5, meaning that the grass cover cannot be reduced by more than half in one time step. The formula for vegetation growth needs to be altered to take into account the impact of consumption on the grass cover. Growth becomes equal to r_V multiplied by VEGPOP times $(1 - \text{XSUTIL})$, multiplied in turn by $\{1 - (\text{VEGPOP times}\,(1 - \text{XSUTIL})\}$, that is, the amount of vegetation in the logistic equation is replaced by the reduced amount following consumption. The vegetation population (cover) the next year is then equal to the cover the previous year, reduced by the $(1 - \text{XSUTIL})$ factor, and incremented by VEGGROW. It needs to be bounded at some nonzero minimum, say 0.2. Copy these formulae for the vegetation dynamics down, and note how the vegetation cover and production relative to consumption changes as the white rhino population grows toward its carrying capacity. Depending on the values you set for F_V and r_V, there may be no interaction (the overgrazing threshold is not exceeded), persistent oscillations in herbivore and vegetation biomass, or a crash toward the minimum vegetation cover and a greatly reduced white rhino population. With K for the white rhino population set at $10,000 \, \text{kg km}^{-2}$, start with values for r_V around 1.0 and F_V around $500 \, \text{kg ha}^{-1}$. Extend the time span if necessary to observe the pattern.

If you encounter difficulties in interpreting the correct formulae, the spreadsheet functions in each cell are displayed in Table 11A.1.

c. Adding variable rainfall

To contain the rainfall data, an additional column needs to be inserted after YEAR. In this column, enter a sequence of annual rainfall totals, calculated relative to some mean value. Here are some real rainfall figures derived from the Kruger Park headquarters at Skukuza, duplicated at the end to span a 100-year period:

1, 0.92, 1.03, 0.69, 1.2, 0.81, 2.03, 0.77, 0.53, 0.88, 1.02, 1.16, 1.1, 0.86, 0.93, 0.88, 0.97, 0.94, 1.07, 0.88, 1.49, 1.33, 0.88, 1.16, 1.01, 0.95, 0.8, 0.93, 0.67, 0.98, 0.98, 1.03, 0.73, 0.73, 1.1, 0.89, 1.1, 1.13, 0.9, 1.3, 0.96, 1.12, 1.05,

0.8, 0.88, 0.64, 0.74, 0.83, 1.15, 0.75, 0.98, 0.69, 0.92, 1.56, 0.59, 1.47, 1.28, 1.22, 1.29, 1.41, 0.87, 1.01, 1.42, 0.79, 0.49, 0.87, 1.43, 0.87, 0.67, 1.25, 12.0, 0.91, 0.79, 0.43, 1.11, 0.65, 1.20, 1.67, 0.99, 0.91, 1.47, 0.92, 1.03, 0.69, 1.2, 0.81, 2.03, 0.77, 0.53, 0.88, 1.02, 1.16, 1.1, 0.86, 0.93, 0.88, 0.97, 0.94, 1.07, 0.88, 1.49

The value for VEGPROD needs to be multiplied additionally by this factor, that is, if the rainfall is higher the grass population produces more forage for consumption by white rhinos, and vice versa. This affects whether the overgrazing threshold is exceeded. Copy this modification down, and note what difference it makes to the graphical output.

d. Incorporating dispersal movements

This entails adding an additional loss function, over and above mortality, to those population stages dispersing, that is, subadults. Since the spreadsheet formulae are already quite complicated, I suggest that you explore the effects of dispersal using the program RHINODYN.

Appendix 12.1   Formulating a model of the seasonal biomass dynamics of a grass forage resource under grazing

Start by retrieving the model developed for the interactive dynamics of herbivores and vegetation in Appendix 5.2. Alter the column heading "TIME" to "WEEKS," and insert a column headed "SEASON." Season gets a value "1" if the week falls during the growing season, and "0" if it falls during the dormant season. Assuming a 6-month growing (summer or wet) season followed by a 6-month dormant (winter or dry) season, enter the value "1" for the first 26 weeks followed by "0" for the next 26 weeks. Extend the time down to 104 weeks so that it spans 2 years, repeating the same seasonal pattern. Change the column heading "VEGET" to "VEG-BIOM," then insert a column headed "VGROW." Start with a vegetation biomass of $25 \, \mathrm{g \, m^{-2}}$ in week 0, then enter the logistic growth equation to calculate the weekly growth increment in the "Vgrow" column. Retain a setting of $500 \, \mathrm{g \, m^{-2}}$ for the saturation biomass of grass, and 0.5 for the maximum growth rate of the grass, now on a weekly basis, but additionally multiply the grass growth rate by the number in the "Season" column. This ensures that growth occurs only during the growing season weeks, and becomes zero during the dormant season.

To allow for the death or turnover of green leaves during the course of the growing season, insert another column headed "VDEATH." The amount of grass dying each week is the green biomass multiplied by the proportional death rate, set at the top of this column to be 0.1 per week. Then insert a column headed "VNECROM," to contain the accumulating standing dead (brown) tissues or necromass. The increment in necromass from grass dying must be counterbalanced by the decay of this dead material, which occurs rapidly during the growing season but much more slowly during the dormant season. Hence two decay rates need to be specified, a growing season decay set at 0.2 per week and a dormant season decay set at 0.02 per week. The growing season decay needs to be multiplied by the value for season, so that it is switched off during the dormant season. Insert a further column "VTOTAL" for the total amount of vegetation, which is the sum of the biomass and the necromass. Two further small modifications are needed. In the logistic formula for vegetation growth, the biomass value restricting the growth rate relative to the saturation level K_V should be the total amount of vegetation ("Vtotal") rather than just the amount of green biomass. A minimum level of green biomass initiating growth at the start of each growing season needs to be specified, using a MAX function. The logical contingency is this: if the calculated amount of green biomass is less than this level, the amount of green biomass becomes this minimum amount, multiplied by the value of "season." This allows the green biomass to drop below this minimum during the dormant level. The minimum level should be set at $25 \, \mathrm{g \, m^{-2}}$ at the top of the worksheet.

You have now established a basic model for seasonal grass dynamics. Copy the formulae down over the 104-week period. Keep the herbivore biomass at zero, so that the pattern of grass dynamics in the absence of grazing can be observed. Plot both the green biomass and the total amount of green plus dead biomass against time in weeks.

To observe the effect of herbivory, the intake rate response in the column headed "CONSUM" needs to be modified in two ways. First, the parameter values set up previously for an annual time step need to be altered to weekly rates. In the hyperbolic formula $I = i_{max}(F/(b + F))$, set $i_{max} = 0.18$ per unit of herbivore biomass per week (equivalent to a maximum consumption rate of 2.5% of body mass per day). Make the half-saturation level $b = 20 \, \mathrm{g \, m^{-2}}$. Since herbivores select green leaves, the effective amount of food available F should have the necromass component down-weighted. Let $F =$ (green biomass) $+ 0.5$ (brown necromass). In converting this intake into herbivore biomass, the effective food quality c needs to be adjusted for the relative proportions of green and brown leaves consumed. Set $c = 0.6$ for green leaves and 0.4 for brown leaves. Insert a column headed "FQUAL," and enter a formula multiplying the conversion coefficient c for green leaves by the proportion of green leaves consumed plus the conversion coefficient for brown leaves times the conversion coefficient for brown leaves, remembering to down-weight the brown leaf amount by a factor of two. The herbivore biomass, in grams per square meter, then has added to it each week the amount of vegetation consumed times the effective quality of this food. For the metabolic maintenance cost m subtracted, set m to be 0.08 per unit of herbivore biomass per week. Hence the maximum consumption rate is just over twice the metabolic maintenance cost.

Last, allowance must be made for the effect of consumption on grass dynamics. The basic principle is that grass grows back *after* consumption has occurred. Hence the grass biomass is equal to the biomass in the previous week, less the amount consumed, plus regrowth. Furthermore, consumption has to be partitioned between the amount removed from green biomass and the amount removed from brown necromass. The grass biomass in the formulae calculating growth, death, and change in necromass must also be adjusted to incorporate the amount remaining *after* consumption. If you have difficulty interpreting these guidelines, refer to Table 12A.1, which shows the spreadsheet formulae that should appear in the respective columns.

Now introduce a nonzero herbivore biomass and establish the stocking level that can be supported by the grass resource, varying seasonally both in amount and in quality. The herbivore biomass will increase during the vegetation growing season and shrink during the dormant season, in response to changing food quality. With the above parameter settings, if you start with an initial herbivore biomass under $50\,\mathrm{g\,m^{-2}}$, the animal biomass at the end of the 2 years should be greater than the initial amount, that is, there is a productive surplus to sell. If you exceed this stocking level initially, the herbivore biomass will end up lower, representing overstocking. A herbivore biomass of $50\,\mathrm{g\,m^{-2}}$ converts to $500\,\mathrm{kg\,ha^{-1}}$, equivalent to a stocking rate of one large cow per hectare, which is not unrealistic.

The last modification to introduce is to allow the seasonal alternation to be less regular. In particular, make the occurrence of rainfall enabling grass growth a random variable with different probability levels during the wet and dry season weeks. Copy the previous model into a new worksheet. Set the probability level for rainfall during the first 26 weeks of each annual cycle at 0.8, and that during the dry season at 0.2. In the "season" column, set the contingency such that if RAND() is less than the applicable probability for that week, season gets the value 1, else it is 0. Extend the period covered to 5 years, rather than just 2. You should find that the outcome from a particular stocking level becomes uncertain so that it is difficult to identify what herbivore biomass can be sustained in the long term.

Table 12A.1 Spreadsheet formulae for the seasonal grass growth dynamics model under grazing

A	B	C	D	E	F
WEEKS	SEASON	VEGBIOM	VGROW	VDEATH	VEGNECR
A0 + 1	1	MAX(C10 − I10*C10/(C10 + F10/2)+D10−E10,C$4*B11)	C11*B11*D$2*(1 − G11/C$2)	C11*E$2	F10 − I10*(F10/2)/(C10 + F10/2) + E10 − (F10 − I10*(F10/2)/(C10 + F10/2))*F$2 − (F10 − I10*(F10/2)/(C10 + F10/2))*F$3*B10

G	H	I	J
VTOTAL	HERBIV	CONSUM	FQUAL
C11 + F11	H10 + J10*I10 − I$1*H10	H11*I$0*(C11 + F11/2)/(I$3 + (C11 + F11/2))	I$1*C11/(C11 + F11/2) + J$1*(F11/2)/(C11 + F11/2)

Appendix 12.2   Spreadsheet model for harvesting a tropical hardwood forest

The spreadsheet representation of a matrix transition model requires columns to represent the stage classes distinguished, and rows above to contain the rate parameters for each stage. Set up columns headed "TIME" and SD1 to SD6 to present the stem diameter classes distinguished, followed by columns headed "TOTALN," "BIOMASS," and "GROWTHRATE." Above these column headings, set aside four rows designated "Ingrowth," "Persist," "Transit," and "Size." The parameter values for each size class are listed in Table 12.1. The first row gets the values for the ingrowth contribution per tree. The second row get the values for the proportion persisting in the same size class. The third row gets the values for the proportion moving upward into the next size class. The fourth row gets the value for the midpoint of the stem diameter range for the class. At time zero, enter the initial numbers observed in each size range in 1956, converted to stems per hectare. Time is incremented in steps of 18 years, to correspond with the period over which the rate parameters were estimated. In the first time step beyond zero, enter formulae to calculate the number of stems entering or persisting in each size class, recalling the models that you developed for Chapter 4. For example, the number of stems in size class SD1 is the sum of the ingrowth contributions from all mature classes, plus the number of stems persisting in this size. The number in size class SD2 is the number entering from size class SD1, plus the number still persisting in this size, and so on. The "TotalN" is simply the total number of plants summed across all size classes. The biomass is calculated by multiplying the number in each size class by the midpoint stem diameter cubed, as an approximation of relative wood volume and hence mass, and then totaling the contributions. Divide the total by 1,000 so that the number for biomass is not too large to be displayed. Calculate the growth rate as the proportional biomass increment between one time step and the next. Then copy these formulae down over a time range spanning at least 300 years as an appropriate horizon for planning the harvests of trees potentially living this long. Observe how the biomass growth rate eventually stabilizes at a steady decline of -0.047 per year if you have formulated the model correctly. Plot the changes in both total plant number and biomass against time, noting how stem number declines progressively while biomass first increases and then declines.

To estimate the potential growth rate, copy the model into a new worksheet. Assume that crowding affects only growth between size classes and not overall survival of a class. Make the value for persistence zero except for the last class SD6. Set the values of the transition rates equal to the survival rates, which is the sum of the persistence in and transition rate from a class (see Table 12.1). Start with an initial cohort of 1,000 plants in class SD1, and none in any of the larger classes, that is, the situation after clear-felling. Note how the growth rate eventually stabilizes at just over 40% per 18 year step, although it takes over 300 years for this to occur. Both biomass and stem number increase exponentially in tandem.

To enable harvesting, density-dependent declines in the transition rates, between size classes as the population biomass increases need to be incorporated. Linear relationships will be assumed, that is, $X = a + bY$, where X is the transition rate, Y the total tree biomass, a the transition rate when biomass is close to zero, assumed equal to the survival rate for the class, and b the slope of the linear decline in transition rate with increasing biomass, which is effectively equal to the value for the persistence rate in a size class when the biomass increase rate becomes zero. Thus, as biomass grows, an increasing proportion of trees remains in the same size class one time step later, while a decreasing proportion grows into the next larger class. Change the row labeled "Persist" to "Slope," and the row labeled "Transit" to "Survive." New values representing overall survival need to be entered in the latter row. The persistence rate in class SD6 also needs to be moved down one row because it effectively represents the survival rate in this class. Table 12A.2 tabulates the formulae that should appear in the row for the first time step. A value is also required for the biomass level at which the biomass increase approaches zero. Enter the value 12,000, which is slightly above that calculated from the observed population structure at time zero. The actual total biomass will stabilize a little below this figure because the observed survival and slope parameters represent a population in slow decline. The tree biomass grows initially and then declines as the size structure stabilizes.

To ascertain the best harvesting policy, first remove all the trees in the designated size range at time zero, and then observe the subsequent time at which the annual biomass increment, and mean increment over the period since harvesting, are greatest. This will require two further columns to calculate these measures. If only class SD6 is harvested, the regrowth increment should appear highest at the next time step, and decline thereafter. If classes SD4 through SD6 are harvested, the regrowth increment will peak two time steps, that is, 36 years later. If all size classes above SD1 are removed, regrowth will only be maximized 4 time steps or 72 years later. Note that with the coarse 18-year time steps, the period when the current increment and mean increment since harvesting are maximal cannot be distinguished.

Copy the model into a new worksheet and apply these harvesting regimes. For example, remove all trees appearing in size class SD6 in every

Table 12A.2 Spreadsheet formulae for forest tree dynamics model

	A	B	C	D	E	F
	TIME	S1	S2	S3	S4	S5
18		B10*(B$9 – (B$9 – B$10*I10/I$11)) + D10*D$8 + E10*E$8 + F10*F$8 + G10*G$8	C10*(C$9 – (C$9 – C$10*$I10/$I$11)) + B10*(B$9 – B$10*$I10/$I$11)	D10*(D$9 – (D$9 – D$10*$I10/$I$11)) + C10*(C$9 – C$10*$I10/$I$11)	E10*(E$9 – (E$9 – E$10*$I10/$I$11)) + D10*(D$9 – D$10*$I10/$I$11)	F10*(F$9 – (F$9 – F$10*$I10/$I$11)) + E10*(E$9 – E$10*$I10/$I$11)

	G	H	I
	S6	TOTAL	Biomass
	G10*(G$9) + F10*(F$9 – F$10*$I10/$I$11)	SUM(B11:G11)	(G11*G$11^3 + F11*F$11^3 + E11*E$11^3 + D11*D$11^3 + C11*C$11^3 + B11*B$11^3)/1000

	J	K	L
	Bgrowth	BiomIncrt	MeanIncrt
	I12/I11 – 1	I12 – I11	AVERAGE (K5:K11)

time step. The removal can be represented by adjusting the formulae so that no tree persists in this size class from one time step to the next. Calculate the harvest obtained as biomass by multiplying the number of trees removed by the stem diameter cubed, divided by 1,000 for convenience. Sum these harvests over an approximately 300-year period and work out the mean annual yield. Then remove all trees in size classes SD4, SD5, and SD6 in every second time step. Last, follow a similar procedure removing all trees larger than size SD1 in every fourth time step. Assess which harvesting size range and felling interval gives the greatest annual yield, and observe the effect on the tree population size and biomass.

Appendix 13.1   Formulating a Markovian transition model for vegetation succession

a. Forest succession

Start by setting up a worksheet with an appropriate heading and space for fixed parameters, as usual. Head four columns "HERBS," "PIONEER," "MIXED," and "CLIMAX," representing herbaceous meadow, pioneer trees, mixed woodland, and climax forest, respectively. Above the headings enter the respective transition probabilities between these states, in successive lines labeled at the left with the end state. In the column "Herbs" enter the following values: 0.1, 0.85, 0.05, 0. This means that there is a 10% chance of the herbaceous state persisting as herbaceous one time step later, an 85% chance that the herbaceous state will be transformed into pioneer woodland, a 5% chance that it will have been replaced by mixed woodland, and zero chance that it will have become climax forest. Consider the time step as being about a decade. Note how the set of probabilities adds up to one. For transitions from the pioneer woodland state, enter these values: 0.05, 0.5, 0.4, 0.05. The 5% chance that the pioneer woodland will revert back to the herbaceous state allows for occasional disturbances like a windstorm followed by a fire removing dead wood. This could represent 5% of the patch area being transformed every decade, or complete transformation once every 20 decades. For transitions from mixed woodland, enter the values 0.05, 0, 0.45, 0.5. Finally, for transitions from climax forest enter the values 0.05, 0, 0, 0.95. This means that climax forest patches remain in this state, unless blown down by a windstorm or similar extreme event occurring once every 200 years on average.

Start at time zero with the entire landscape covered by herbaceous vegetation, that is, representing the situation immediately following a major disturbance. Hence enter 1 in the herbaceous column and 0s in the three

Table 13A.1   Spreadsheet formulae for calculating state transitions for the forest succession model

A	B	C	D	E	F
TIME	HERBS	PIONEER	MIXED	CLIMAX	SUM
A20+	B20*B$10+	B20*B$11+	B20*B$12+	B20*B$13+	SUM
1	C20*C$10+	C20*C$11+	C20*C$12+	C20*C$13+	(B21:E21)
	D20*D$10+	D20*D$11+	D20*D$12+	D20*D$13+	
	E20*E$10	E20*E$11	E20*E$12	E20*E$13	

woodland columns. Add an additional column "SUM" to add up the proportions in each state. These proportions must sum to 1.0, which is a check that the transition probabilities from any state likewise add up to one.

In the next line, enter the formulae needed to calculate the landscape proportion in each state in the following time step, based on the previous proportions and the transition probabilities. This is obtained by multiplying the prior proportion in each state by the respective transition probability into the particular end state represented by the column, summed over all states (see Table 13A.1). Copy these formulae down over a sufficient time period for the proportions to stabilize.

The questions to be addressed are (i) how long will the area take to recover to prevalently climax woodland and (ii) what will be the proportions of the different states at this end stage? Based on the above transition probabilities, recovery toward predominantly climax forest should take around 20 decades. The proportion of the landscape covered by vegetation other than climax forest will still remain over 20%. Graph the changing proportions of each successional stage over time. Experiment by changing the transition probabilities to values you consider more realistic, and note the outcome for the eventual landscape composition.

b.   River succession

For successional changes taking place in a river channel, the basic model remains unchanged; only column headings and the associated transition probabilities need to be changed. Hence copy the previous model into a new worksheet, and type these column headings: "WATER," "SAND," "REEDS," and "FOREST." Enter the respective transition probabilities between these patch states, given in Table 13.2a, at the top. Start with most of the river channel being either water or sand, and only small proportions of reeds and forest, representing the situation after a major flood. Within less than a century, most of the riparian zone will have become vegetated by either reeds or forest, and the forest will continue to expand until it fills most of the channel. This is because there was only a small chance of forest patches reverting to water or sand, as a result of low intensity floods.

This simple model fails to take into account the massive changes that are brought about by major floods. These floods occur irregularly at long intervals. They represent stochastic events, and hence the model needs to be made stochastic to incorporate their effects. The state transitions following such events differ radically from the slow transformations that take place between them. When a major flood occurs, a substantial proportion of the forest and some of the reed-beds get washed away. Hence the model needs to represent two sets of transition probabilities, one under normal conditions and another contingent upon the occurrence of a big flood.

Accordingly, copy the deterministic river succession model into a new worksheet. Add two further columns, one headed "RAND" and the other "FLOOD." In the column headed "Rand," enter the formula for retrieving a randomly generated number between 0 and 1, that is, =**RAND**(). Above it, enter the probability value for a big flood as 0.1, that is, on average these events occur once in a century. Under the heading "Flood," enter the contingent formula, IF the value of the random number is LESS than the flood probability, the text "FLOOD" appears, otherwise the cell remains blank. Then enter two sets of transition probabilities, one for normal decades and another for decades when a major flood takes place. Suggested values are listed in Table 13.2b and c. The formulae for calculating the distribution of states in each successive decade also needs to be changed to reflect the flood contingency, that is, if no flood occurs, the first set of transition probabilities applies, otherwise the second set governs transitions. If you have difficulty in entering the appropriate functions, refer to Table 13A.2. Graph the changes in river state over 50–100 decades, and note how the composition of the river channel varies quite widely between extremely open or wooded each time you press F9 to generate a new random number sequence. Sometimes two big floods occur in close succession, while in other trials there may be an extended period without floods, allowing forests to become prevalent.

Appendix 13.2   Savanna management model

In a new worksheet, set up columns to represent three savanna patch states: parkland with tall grass, open woodland with short grass lawns, and woody thicket. Table 13.3 presents suggested values for transition rates between these states, under average rainfall conditions. Start with the area covered

Table 13A.2 Spreadsheet formulae for river succession model incorporating stochastic floods

A	B	C	C	E
TIME	WATER	SAND	REEDS	FOREST
A20 + 1	IF(G20 > G$19, B20*B$15 + C20*C$15 + D20*D$15 + E20*E$15, B20*G$15 + C20*H$15 + D20*I$15 + E20*J$15)	IF(G20 > G$19, B20*B$16 + C20*C$16 + D20*D$16 + E20*E$16, B20*G$16 + C20*H$16 + D20*I$16 + E20*J$16)	IF(G20 > G$19, B20*B$17 + C20*C$17 + D20*D$17 + E20*E$17, B20*G$17 + C20*H$17 + D20*I$17 + E20*J$17)	IF(G20 > G$19, B20*B$18 + C20*C$18 + D20*D$18 + E20*E$18, B20*G$18 + C20*H$18 + D20*I$18 + E20*J$18)

F	G	H
SUM	RAND	FLOOD
SUM(B21:E21)	RAND()	IF(G21 > G$19, "", "FLOOD")

entirely by parkland, and note the eventual distribution of patch states projected by this set of transition probabilities. You should find that about 75% of the landscape supports parkland, 15% thicket, and under 10% lawn. Graph the change in these proportions over time. This represents the ungrazed situation with rainfall constant at average levels.

Next allow for variable rainfall. Copy the model into a new worksheet, and insert two extra columns, one headed "RAND" and the other "RAIN." For simplicity, categorize the rainfall as either average, high, or low. Assume that average rainfall is expected in half the years, high rainfall in a quarter, and low rainfall in the remaining quarter. However, rainfall is a stochastic variable, so you need to call upon the random number generator to decide what rainfall occurs in any particular year. Hence in the column headed "Rand," enter the function =**RAND**() and copy it down over 30 rows to generate a sequence of random numbers. In the column headed "Rain," enter the following formula:

$$= \textbf{IF(B16} < \textbf{0.25, "LOW," IF(B16} > \textbf{0.75, "HIGH," "AVER"))}$$

This means that if the value of the random number in column B is less than 0.25, the text "LOW" will appear, while if it is greater than 0.75, the text "HIGH" will appear, else "AVER" will be printed.

Now three sets of transition probabilities are needed, depending on whether the rainfall is low, average, or high. Refer to Table 13.3 for suggested values, and enter them above the column headings in the appropriate columns. The parkland state has a greater chance of persisting when the rainfall is high, and fires consequently hot, and these hot fires are more effective in reducing the extent of thicket. Under low rainfall conditions, lawn and thicket are more likely to persist or expand at the expense of parkland. Next, the formulae calculating the patch state distribution one year later need to be modified to incorporate the rainfall dependency. In concept, if rainfall is average, one set of transition rates applies; if it is above average, a second set gets used; if it is below average, a third set must be used. If you have difficulty interpreting this, refer to Table 13A.3. Copy the formulae down and adjust the graph to reflect the values in the current worksheet. Observe how the patch state proportions fluctuate, depending on the rainfall. During periods of low rainfall, the parkland state shrinks in its distribution, while thicket and lawn expand, and vice versa. Since the effect of low rainfall is greater than that of high rainfall, the mean extent of parkland tends to be less than that for constant average rainfall, but the actual proportions over any period can change quite greatly from one trial to the next if the F9 button is pressed repeatedly. Fortuitous periods with low rainfall in several successive years can lead to a great expansion in thicket at the expense of parkland. Similarly, extended periods of high rainfall can bring about a high predominance of parkland.

Third, bring in the additional effect of stocking with cattle on the transition probabilities. Copy the model into a new worksheet, and label an additional column "STOCKDENS." Again, set the stocking density simply as either low, medium, or high, symbolized by the letters "L," "M," or "H" in this column. Start with low stocking by entering a series of Ls. The effect of grazing is to promote the expansion of short grass lawns at the expense of tall grassland. Specifically, assume that low stocking increases the persistence or expansion of lawns by 0.05, medium stocking increases this by 0.1, and heavy stocking raises it by 0.2. This is most easily handled by entering these values in a new column headed "EFFECT," contingent upon the letter appearing in the column headed "Stockdens." The formulae governing the transitions between patch states then need to be modified to reflect the effect of grazing, done simply by (i) adding the value in the "Effect" column to the rainfall-dependent value of the transition probability for transitions toward lawn, and (ii) subtracting the same value from transition probabilities toward parkland. However, in cases where the subtraction would produce a negative number, the number added and subtracted must be adjusted to avoid this. Furthermore, with moderate or heavy stocking under conditions of low rainfall, thicket rather than lawn is likely to expand at the expense of transitions toward parkland. Table 13A.4 presents the spreadsheet formulae containing these small adjustments, if you have difficulty interpreting them.

The question to be addressed is, what stocking density is optimal? Change the stocking density to "M," and observe how this causes a reduction in the extent of parkland in favor of lawn. Elevate it further to "H," and note how parkland shrinks further while thicket expands, especially if a sequence of low rainfall years is experienced.

The vegetation impact is obvious, but what difference does this make to the economic profitability? To assess this, assume that profitability depends on (i) the stocking density, (ii) the extent of parkland which provides most grazing, and (iii) the extent of thicket, which offers no useful forage. Furthermore, although high stocking is indexed in the "Effect" column as having twice the weight as medium stocking, the corresponding profit will not be twice as high because the additional animals grow less well and produce lower quality meat at high stocking. A square-root transformation lowers the effective difference in value between the three stocking densities. Hence in a further column headed "PROFIT," enter this formula:

$$= (I16)^{\wedge}0.5{*}10{*}D16{*}(1 - F16)$$

Table 13A.3 Spreadsheet formulae for savanna patch dynamics model incorporating variable rainfall

A	B	C	D	E
TIME	RAND	RAIN	PARKLAND	LAWN
A20+	RAND()	IF(B21 < 0.25, "LOW",	IF(B20 > 0.75,	IF(B20 > 0.75,
1		IF(B21 > 0.75, "HIGH",	D20*D$11+E20*E$11+F20*F$11,	D20*D$12+E20*E$12+F20*F$12,
		"AVER"))	IF(B20 < 0.25,	IF(B20 < 0.25,
			D20*D$5 + E20*E$5 + F20*F$5,	D20*D$6 + E20*E$6 + F20*F$6,
			D20*D$8 + E20*E$8 + F20*F$8))	D20*D$9 + E20*E$9 + F20*F$9))

F	G
THICKET	SUM
IF(B20 > 0.75,	SUM (C21:F21)
D20*D$13+E20*E$13+F20*F$13,	
IF(B20 < 0.25,	
D20*D$7 + E20*E$7 + F20*F$7,	
D20*D$10 + E20*E$10 +	
F20*F$10))	

where column I contains the values for "Effect," column D the proportion of parkland, and column F the proportion of thicket. Copy it down, and then in another cell enter a function to calculate the average profit over some extended period to compare the yields from different stocking densities. Leave out the initial five rows, which simply reflect the starting vegetation condition. First, compare a stocking level maintained consistently at "medium" with returns from either "low" or "high" stocking. You should find that medium tends to be best on average. However, returns from high stocking vary quite widely depending on the rainfall, and sometimes exceed those from medium stocking. Thereafter, consider also the results from a variable stocking strategy, contingent upon the rainfall received in the preceding year. To do this, set the stocking level initially at "low" across all years. Then examine the actual rainfall in each successive year and adjust the stocking level up to either medium or high as you see fit. Assess how much improvement such adaptive management makes to the overall profitability.

Appendix 14.1   Habitat occupation model for overwintering geese

a. Basic model

This simple model considers the effects of weekly changes in food abundance on food intake rate across three available habitats. Geese selectively occupy the habitat yielding highest gains during each week. Food abundance is reduced solely through consumption. There is no regeneration or intrinsic decay of food, nor any effect of interference on food intake.

Underneath the headings and space for the parameter values, set up the first column to show time incremented in weeks. The next six columns represent the weekly food abundance and corresponding daily rate of food intake per unit mass of geese for the three habitats labeled A, B, and C. "A" represents the natural habitat potentially yielding highest food gains. "B" represents an alternative habitat such as agricultural fields that is at best barely adequate for population maintenance. "C" represents a buffer habitat yielding submaintenance gains. The food intake rate follows a Michaelis–Menten or "Type II" functional response, that is, $i_{max}F/(f_{1/2} + F)$, where F represents food availability in mass per unit area, i_{max} the maximum food intake rate at high food abundance, and $f_{1/2}$ the food abundance level at which the intake rate drops to half of its maximum. Suggested parameter values for the three habitats are listed in Table 14A.1. Enter these values above the respective column headings.

Table 13A.4 Spreadsheet formulae for savanna patch dynamics model incorporating variable rainfall and different stocking densities of grazers

A	B	C	D	E
TIME	RAND	RAIN	PARKLAND	LAWN
A20 + 1	RAND()	IF(B21 < 0.25, "LOW," IF(B21 > 0.75, "HIGH," "AVER"))	IF(B20 > 0.75, D20*(D$11 − I20) + E20*(E$11 − I20) + F20*(F$11 − I20), IF(B20 < 0.25, D20*(D$5 − I20) + E20*(E$5 − 0.1) + F20*F$5, D20*(D$8 − I20) + E20*(E$8 − I20) + F20*(F$8 − I20)))	IF(B20 > 0.75, D20*(D$12 + I20) + E20*(E$12 + I20) + F20*(F$12 + I20), IF(B20 < 0.25, D20*(D$6 + I20) + E20*(E$6) + F20*F$6, D20*(D$9 + I20) + E20*(E$9 + I20) + F20*(F$9 + I20)))

F	G	H	I	J
THICKET	SUM	STOCK DENS	EFFECT	PROFIT
IF(B20 > 0.75, D20*D$13 + E20*E$13 + F20*F$13, IF(B20 < 0.25, D20*D$7 + E20*(E$7 + 0.1) + F20*F$7, D20*D$10 + E20*E$10 + F20*F$10))	SUM (C21:F21)	L	IF(H21 = "L," 0.05, IF(H21 = "M," 0.1, 0.2))	(I21)^0.5*10* D21*(1 − F21)

In an additional line, specify the relative extent of the three habitats. Initially set this value to 1.0 for each habitat, that is, they occupy equivalent areas. Head the next three columns "CHOICE", with a subheading for the three habitats A, B, and C. The following two columns are headed "POP" and "GAIN," respectively. At time zero, set the initial food abundance to be 100 units (in $g\,m^{-2}$, or $kg\,ha^{-1}$, depending on what is appropriate for the goose species concerned). For each habitat type, enter the functions to calculate the respective food intake rates, in a spreadsheet version of the Michaelis–Menten equation specified above.

The columns headed "Choice" reflect the chosen habitat each week, that is, that yielding the highest food intake rate. This represents a contagious distribution, with the entire goose population aggregating in the best habitat. Thus, each cell receives either a 1 or 0, and no more than one column can contain a 1. To achieve this, the following contingent function should be entered in the choice cell for habitat A at time zero:

$$= \mathbf{IF(C15 < E15, 0, (IF(G15 > E15, 0, 1))).}$$

This means that if the food intake rate obtained from habitat A (contained in cell C15) is less than that from habitat B (in cell E15), this cell receives the value 0, otherwise it gets 1, provided the food intake rate obtained from habitat C (in cell G15) is not greater than that from habitat B. The required choice function for habitat B is

$$= \mathbf{IF(E15 < C15, 0, (IF(E15 > G15, 1, 0))),}$$

and that for habitat C is

$$= \mathbf{IF(G15 > E15, 1, 0).}$$

These equations achieve the required outcome, that is, just one of the three columns receives the value 1 each week.

Table 14A.1 Suggested parameter values for three habitat types available to a hypothetical goose population

	Habitat A	Habitat B	Habitat C
Initial food abundance F in mass per unit area	100	100	100
Maximum intake rate i_{max} as a proportion of goose biomass	0.18	0.12	0.08
Half-saturation level $f_{1/2}$ in mass per unit area	8	10	12

Initially set some arbitrary value for the population size, expressed as a biomass density (amounting to less than 10% of the food biomass). The column headed "Gain" receives the food intake rate from the chosen habitat. The formula to achieve this is

$$= C15^*H15 + E15^*I15 + G15^*J15$$

that is, each intake rate is multiplied by the respective choice value for that habitat.

In week 1, a formula is needed to calculate the food abundance remaining in each habitat, following consumption. For habitat A, this is

$$= IF(B15 - (C15^*H15^*K15^*7)/C\$12 < 0, 0,$$

$$B15 - (C15^*H15^*K15^*7)/C\$12).$$

The first contingency ensures that the value for food abundance cannot drop below zero. If this is not the case, the food abundance remaining is calculated by subtracting the product of the relative intake rate, choice, and population biomass, multiplied by seven to convert from daily to weekly consumption. Corresponding formulae should be entered for the other two habitats. The functions can then be copied down over the 26 weeks representing the overwintering period. Graphing the changes in habitat use through the winter (i.e., the values in the three choice columns) is helpful for depicting the outcome.

With the suggested parameter values, the goose population will occupy solely habitat A throughout the winter period provided that its biomass level does not exceed three units ($g\,m^{-2}$ or $kg\,ha^{-1}$, depending on the units chosen for the food abundance). At higher abundance levels, the population shifts across to habitat B partway through the winter, perhaps with brief return forays to habitat A. Above eight units the population also uses habitat C toward the end of the winter. Hence this model suggests the extent of the problem likely to be encountered by farmers as a result of the geese feeding in their fields or arable lands after the goose population has grown.

b. Incorporating goose population dynamics

To incorporate changes in the abundance of the goose population as a result of the food consumed, the nutritional gain over and above maintenance needs to be translated into a corresponding rate of population increase, counterbalancing overwinter losses against the reproductive increment during the breeding season. Hence add a further column headed "POP-GROW." This function needs parameter values for the conversion factor from food eaten into goose biomass, and for the metabolic maintenance cost in similar units. The formula is

$$= (L15^*M\$11 - M\$12)^*0.1 + 1$$

where cell M11 contains the conversion coefficient (suggested value 0.9) and cell M12 the metabolic cost (suggested value 0.1). In practice allowance should also be made for the proportion of the 24-h day that the geese spend foraging, but we will ignore this constant. The factor 0.1 converts to reasonable units for a weekly proportional mass gain, expressed as a factor by adding 1 to it. This represents the weekly contribution to the population growth or decline over the 26 weeks. The overall outcome is obtained as the product of these 26 weekly values. With a low population level, the maximum population growth rate is just under 14% per year for the chosen parameter values, which seems reasonable (allowing for some inevitable mortality among adults). The net population growth becomes negative if the population abundance much exceeds 10 units.

Now you can explore the potential effects of changes in habitat features on the goose population. If the extent of the most favorable habitat A is halved (by changing the setting for "Area" from 1.0 to 0.5), the zero growth level for the geese is lowered by a third. An equivalent reduction in the extent of habitat B reduces the goose abundance that can be sustained by only 15%, while eliminating the buffer habitat C entirely (by setting its initial abundance to zero) makes very little difference to the goose abundance. However, if access to habitat B is mostly excluded, the goose population supported is lowered by 30%. This suggests the contribution that agricultural fields could make toward the abundance of the geese, whatever the cost to the farmers.

The above simulations assumed that the amount of food produced by each habitat remains constant from year to year. In practice, the productivity of salt-marsh grasses could vary from year to year depending on weather conditions. In years when the production of food in the natural habitat is reduced, the geese exploit alternative habitats in the form of agricultural fields earlier and thus more extensively, with a corresponding reduction in their population growth potential. In years of high production, they may make little or no use of habitats beyond the natural salt-marsh. The potential consequences can be explored by varying the initial food abundance in habitat A. In this situation elimination of the habitat C causes the population to decline more in bad years than when this buffer is available, even though the buffer habitat provides submaintenance returns.

c. Further modifications

The weekly iteration is a crude representation of changes in habitat use that may take place on a daily basis. The model can be adjusted easily to represent daily time steps, simply by changing the amount of food subtracted due to consumption in each line, and also altering the values for the contribution to population growth to a daily basis. Obviously, the number of lines has to be extended correspondingly to span the winter period. The difference from this improvement is quite small, except in representing more frequent daily excursions back to habitat A during the course of the winter period.

In some situations it may be costly to shift between habitats because these are far apart. This means that once the favorable habitat is abandoned, there is no return. This can be represented by changing the formula in the choice column for habitat A to read as follows:

$$= IF(I114 = 0, 0, (IF(C15 > E15, 1, 0)))$$

and correspondingly that for habitat B becomes

$$= IF(H15 = 1, 0, (IF(E15 > G15, 1, 0))).$$

Nonreturn to the most favorable habitat makes only a small difference to the population performance for the suggested parameter values.

The model could also be extended to allow for continuing production by the food plants for at least part of the winter period, and perhaps inherent attrition by these plants even in the absence of grazing.

References

Abrams, P.A. and L.R. Ginzburg. 2000. Anomalous predictions of predation: prey dependent, ratio-dependent or neither. *Trends in Ecology and Evolution* 15: 337–41.

Alvarez-Buylla, E.R. 1994. Density dependence and patch dynamics in tropical rain forests: matrix models and applications to a tree species. *The American Naturalist* 143: 155–91.

Anderson, R.M. and R.M. May. 1981. The population dynamics of microparasites and their invertebrate hosts. *Philosophical Transactions of the Royal Society of London* B291: 451–524.

Anderson, R.M. and R.M. May. 1982. Population dynamics of fox rabies in Europe. *Nature* 289: 765–71.

Anderson, D.R., K.P. Burnham, and W.L. Thompson. 2000. Null hypothesis testing: problems, prevalence, and an alternative. *Journal of Wildlife Management* 64: 912–23.

Barlow, N.D. 1996. The ecology of wildlife disease control: simple models revisited. *Journal of Applied Ecology* 33: 303–14.

Baxter, P.W.J. and W.M. Getz. 2005. A model-famed evaluation of elephant effects on tree and fire dynamics in African savannas. *Ecological Applications* 15: 1331–41.

Bayliss, P. 1989. Population dynamics of magpie geese in relation to rainfall and density: implications for harvest models in a fluctuating environment. *Journal of Applied Ecology* 26: 913–23.

Begon, M., C.R. Townsend, and J.L. Harper. 2006. *Ecology. From Individuals to Ecosystems*, 4th edn. Blackwell, Oxford.

Beissinger, S.R. and D. McCullough (eds). 2002. *Population Viability Analysis. Assessing Models for Recovering Endangered Species.* University of Chicago Press, Chicago.

Beissinger, S.R. and M.I. Westphal. 1998. The use of demographic models of population viability in endangered species management. *Journal of Wildlife Management* 62: 821–41.

Bengis, R.G., C.C. Grant, and V. de Vos. 2003. Wildlife diseases and veterinary controls: a savanna ecosystem perspective. In J.T. du Toit, K.H. Rogers, and H.C. Biggs (eds), *The Kruger Experience. Ecology and Management of Savanna Heterogeneity*, pp. 349–69. Island Press, Washington, DC.

Beres, D.L., C.W. Clark, G.L. Swartzman, and A.M. Starfield. 2001. Truth in modeling. *Natural Resource Modelling* 14: 457–63.

Berryman, A.A. 1992. The origins and evolution of predator–prey theory. *Ecology* 73: 1530–5.

Berryman, A., M. Lima, and T. Coulson (in press). A theoretical framework for deciphering the effects of climate on Soay sheep dynamics. *American Naturalist*.

Bjornstad, O.N., B.F. Finkenstadt, and B.T. Grenfell. 2002. Dynamics of measles epidemics: estimating scaling of transmission rates using a time series SIR model. *Ecological Monographs* 72: 169–84.

Boonstra, R. and P.T. Boag. 1987. A test of the Chitty hypothesis: inheritance of life-history traits in meadow voles. *Evolution* 8: 196–219.

Boyce, M.S. 1989. *The Jackson Elk Herd. Intensive Wildlife Management in North America*. Cambridge University Press, Cambridge.

Boyce, M.S. and L.L. McDonald. 1999. Relating populations to habitats using resource selections functions. *Trends in Ecology and Evolution* 14: 268–72.

Brown, P.M. and R. Wu. 2005. Climate and disturbance forcings of episodic tree recruitment in a southwestern ponderosa pine landscape. *Ecology* 86: 3030–8.

Bryant, J.P., P.J. Kuropat, P.B. Reichardt, and T.P. Clausen. 1991. Controls over the allocation of resources by woody plants to chemical antiherbivore defense. In C.T. Robbins and T. Palo (eds), *Plant Chemical Defenses and Mammalian Herbivores*, pp. 83–102. CEC, Boca Raton, Florida.

Burgman, M.A., S. Ferson, and H.R. Akcakaya. 1993. *Risk Assessment in Conservation Biology*. Chapman & Hall, London.

Burnham, K.P. and D.R. Anderson. 2002. *Model Selection and Inference: A Practical Information-theoretic Approach*, 2nd edn. Springer-Verlag, New York.

Case, T.J. 2000. *An Illustrated Guide to Theoretical Ecology*. Oxford University Press, Oxford.

Caswell, H. 2000. Prospective and retrospective perturbation analyses and their use in conservation biology. *Ecology* 81: 619–27.

Caswell, H. 2001. *Matrix Population Models*, 2nd edn. Sinauer, Sunderlands, MA.

Caughley, G. 1976a. Plant-herbivore systems. In R.M. May (ed.), *Theoretical Ecology – Principles and Applications*, pp. 94–113. Blackwell, Oxford.

Caughley, G. 1976b. The elephant problem – an alternative hypothesis. *East African Wildlife Journal* 14: 265–83.

Caughley, G. 1994. Directions in conservation biology. *Journal of Animal Ecology* 63: 215–44.

Chapin, F.S., T.S. Rupp, A.M. Starfield, et al. 2003. Planning for resilience: modeling change in human–fire interactions in Alaskan boreal forests. *Frontiers in Ecology and the Environment* 1: 255–61.

Clark, C.W. 1990. *Mathematical Bioeconomics. The Optimal Management of Renewable Resources*, 2nd edn. Wiley, New York. (Full analysis of economic aspects of harvesting with supporting mathematics.)

Cleaveland, S., G.R. Hess, A.P. Dobson. et al. 2002. The role of pathogens in biological conservation. In P.J. Hudson, A. Rizzoli, B.T. Grenfell, H. Heesterbroek, and A.P. Dobson (eds), *The Ecology of Wildlife Diseases*, pp. 139–50. Oxford University Press, Oxford.

Clifton-Hadley, R.S., J.W. Wilesmith, M.S. Richards, P. Upton, and S. Johnston. 1995. The occurrence of *Mycobacterium bovis* infection in cattle in and around an area subject to extensive badger control. *Epidemiology and Infection* 114: 179–93.

Clutton-Brock, T.H., B.T. Grenfell, T. Coulson, et al. 2004. Population dynamics in Soay sheep. In T. Clutton-Brock and J. Pemberton (eds), *Soay Sheep. Dynamics and Selection in an Island Population*, pp. 52–88. Cambridge University Press, Cambridge, UK.

Coltman, D.W., J.G. Pilkington, J.A. Smith, and J.M. Pemberton. 1999. Parasite-mediated selection against inbred Soay sheep in a free-living island population. *Evolution* 53: 1259–67.

Coughenour, M.B. 1992. Spatial modeling and landscape characterization of an African pastoral ecosystem: a prototype model and its potential use for monitoring drought. In D.H. McKenzie, D.E. Hyatt, and V.J. McDonald (eds), *Ecological Indicators*, vol. 1, pp. 787–810. Elsevier, London.

Coughenour, M.B. and F.J. Singer. 1996. Elk population processes in Yellowstone National Park under the policy of natural regulation. *Ecological Applications* 6: 573–93.

Courchamp, F., T. Clutton-Brock, and B. Grenfell. 1999. Inverse density dependence and the Allee effect. *Trends in Ecology and Evolution* 14: 405–10.

Coulson, T., E.A. Catchpole, S.D. Albon, et al. 2001. Age, sex, density, winter weather, and population crashes in Soay sheep. *Science* 292: 1528–31.

Coulson, T., F. Guinness, J. Pemberton, and T. Clutton-Brock. 2004. The demographic consequences of releasing a population of red deer from culling. *Ecology* 85: 411–22.

Crawley, M.J. 1983. Herbivory: the Dynamics of Animal–Plant Interactions. Studies in *Ecology* Series No. 10. Blackwell, Oxford.

Crawley, M.J. (ed.). 1992. *Natural Enemies. The Population Biology of Predators, Parasites and Diseases*. Blackwell, Oxford.

Crouse, D.T., L.B. Crowder, and H. Caswell. 1987. A stage-structured population model for loggerhead sea turtles and implications for conservation. *Ecology* 68: 1412–23.

Crowder, L.B., D.T. Crouse, S.S. Heppell, and T.H. Martin. 1994. Predicting the impact of turtle excluder devices on loggerhead sea turtle populations. *Ecological Applications* 4: 437–45.

Daszak, P., A.A. Cunningham, and A.D. Hyatt. 2000. Emerging infectious diseases of wildlife – threats to biodiversity and human health. *Science* 287: 443–9.

DeAngelis, D.L., R.A. Goldstein, and R.V. O'Neil. 1975. A model for trophic interaction. *Ecology* 56: 881–92.

DeAngelis, D.L. and L.J. Gross (eds). 1992. *Individual-based Models and Approaches in Ecology*. Chapman & Hall, London.

Dennis, B. and M.L. Taper. 2000. Joint effect of density dependence and rainfall on the abundance of San Joaquin kit fox. *Journal of Wildlife Management* 64: 388–400.

Deshmukh, I.K. 1984. A common relationship between precipitation and grassland peak biomass for east and southern Africa. *African Journal of Ecology* 22: 181–6.

Dobson, A.P. 1995. The ecology and epidemiology of rinderpest virus in Serengeti and Ngorongoro Conservation Area. In A.R.E. Sinclair and P. Arcese (eds), *Serengeti II: Dynamics, Management and Conservation of an Ecosystem*, pp. 485–505. University of Chicago Press, Chicago.

Dobson, A.P. and M. Meagher. 1996. The population dynamics of brucellosis in the Yellowstone National Park. *Ecology* 77: 1026–36.

Durant, D., H. Fritz, S. Blais, and P. Duncan. 2003. The functional response in three species of herbivorous Anatidae: effects of sward height, body mass and bill size. *Journal of Animal Ecology* 72: 220–31.

Edelstein-Kechet, L. 1988. *Mathematical Models in Biology*. Random House, New York.

Euler, D. and M.M.J. Morris. 1984. Simulated population dynamics of the white-tailed deer in an any-deer hunting system. *Ecological Modelling* 24: 281–92.

Fagan, W.F., E. Meir, J. Prendargast, A. Folarin, and P. Kareiva. 2001. Characterizing population vulnerability for 758 species. *Ecology Letters* 4: 132–8.

Fine, P. and J.A. Clarkson. 1982. Measles in England and Wales – I: an analysis of factors underlying seasonal patterns. *International Journal of Epidemiology* 11: 5–14.

Forain, B.D. and D.M. Stafford Smith. 1991. Risk, biology and drought management strategies for cattle stations in central Australia. *Journal of Environmental Management* 33: 17–33.

Ford, E.D. 2000. *Scientific Method for Ecological Research*. Cambridge University Press, Cambridge.

Fowler, C.W. 1981. Density dependence as related to life history strategy. *Ecology* 62: 602–10.

Fowler, C.W. 1987. A review of density dependence in populations of large mammals. In H.H. Genoways (ed.), *Current Mammalogy*, vol. 1, pp. 401–41. Plenum Press, New York.

Gaillard, J.M., M. Festa-Bianchet, and N.G. Yoccoz. 1998. Population dynamics of large herbivores: variable recruitment with constant adult survival. *Trends in Ecology and Evolution* 13: 58–63.

Gascoyne, S.C., A.A. King, A.A. Laurenson, M. Borner, B. Schildger, and J. Banat. 1993. Aspects of rabies infection and control in the conservation of the African wild dog in the Serengeti region, Tanzania. *Onderstepoort Journal of Veterinary Research* 60: 415–20.

Georgiadis, N., M. Hack, and K. Turpin. 2003. The influence of rainfall on zebra population dynamics: implications for management. *Journal of Applied Ecology* 40: 125–36.

Gilgen, D., B.G. Williams, C. MacPhail, et al. 2001. The natural history of HIV/AIDS in a major gold-mining centre in South Africa: results of a biomedical and social survey. *South African Journal of Science* 97: 387–92.

Gillson, L. and K. Lindsay. 2003. Ivory and ecology – changing perspectives on elephant management and the international trade in ivory. *Environmental Science and Policy* 6: 411–9.

Gilpin, M.E. and F.J. Ayala. 1973. Global models of growth and competition. *Proceedings of the National Academy of Science USA* 70: 3590–3.

Gilpin, M.E. and M.E. Soule. 1986. Minimum viable populations: processes of species extinction. In M.E. Soule (ed.), *Conservation Biology*, pp. 19–34. Sinauer, Sunderland, MA.

Ginzburg, L.R. and D.E. Tanneyhill. 1994. Population cycles of forest Lepidoptera: a maternal effect hypothesis. *Journal of Animal Ecology* 63: 79–92.

Gleick, J. 1987. *Chaos. Making a New Science*. Viking, New York.

Goss-Custard, J.D., R.T. Clarke, S.E.A. le V. Durell, R.W.G. Caldow, and B.J. Ens. 1995b. Population consequences of habitat loss and change in wintering migratory birds. II. Model predictions. *Journal of Applied Ecology* 32: 334–48.

Goss-Custard, J.D., R.T. Clarke, K.B. Briggs. et al. 1995a. Population consequences of winter habitat loss in a migratory shorebird. I. Estimating model parameters. *Journal of Applied Ecology* 32: 317–33.

Gotelli, N.J. 1995. *A Primer of Ecology*, 2nd edn. Sinauer, Sunderland, MA.

Gotelli, N.J. 2001. *A Primer of Ecology*, 3rd edn. Sinauer, Sunderland, MA.

Gotelli, N.J. and A.M. Ellison. 2006. Forecasting extinction risk with nonstationary matrix models. *Ecological Applications* 16: 51–61.

Grant, C.C., T. Davidson, P.J. Funston, and D.J. Pienaar. 2002. Challenges faced in the conservation of rare antelope: a case study on the northern basalt plains of the Kruger National Park. *Koedoe* 45: 45–66.

Green, R.E. 2002. Diagnosing causes of population declines and selecting remedial actions. In K. Norris and D.J. Pain (eds), *Conserving Bird Biodiversity*, pp. 139–56. Cambridge University Press, Cambridge.

Grenfell, B.T. and A.P. Dobson (eds). 1995. *Ecology of Infectious Diseases in Natural Populations*. Cambridge University Press, Cambridge.

Grenfell, B.T., M. Lonergam, and J. Harwood. 1992. Quantitative investigations of the epidemiology of phocine distemper virus (PDV) in European common seal populations. *Science of the Total Environment* 115: 31–44.

van Groenendael, J., H. de Kroon, and H. Caswell. 1988. Projection matrices in population biology. *Trends in Ecology and Evolution* 3: 264–9.

Gutierrez, R.J. and S. Harrison. 1996. Applying metapopulation theory to spotted owl management: a history and critique. In D. McCullough (ed.), *Metapopulations and Wildlife Conservation*, pp. 167–86. Island Press, Washington, DC.

Haefner, J.W. 1996. *Modeling Biological Systems. Principles and Applications*. Chapman & Hall, London.

Hanski, I. 1992. Inferences from ecological incidence functions. *American Naturalist* 139: 657–62.

Hanski, I. 1994. Patch occupancy dynamics in fragmented landscapes. *Trends in Ecology and Evolution* 9: 131–95.

Hanski, I. 1997. Metapopulation dynamics. From concepts and observations to predictive models. In I. Hanski and M. Gilpin (eds), *Metapopulation Biology*, pp. 69–91. Academic Press, San Diego, CA.

Hanski, I. 1999. *Metapopulation Ecology*. Oxford University Press, Oxford, UK.

Hanski, I., M. Kuusaari, and M. Nieminen. 1994. Metapopulation structure and migration in the butterfly *Melitaea cinxia*. *Ecology* 75: 747–62.

Harrington, R., N. Owen-Smith, P. Viljoen, H. Biggs, and D. Mason. 1999. Establishing the causes of the roan antelope decline in the Kruger National Park, South Africa. *Biological Conservation* 90: 69–78.

Harrison, S. and A.D. Taylor. 1997. Empirical evidence for metapopulation dynamics. In I. Hanski and M. Gilpin (eds), *Metapopulation Biology*, pp. 27–42. Academic Press, San Diego, CA.

Harwood, J. and K. Stokes. 2002. Coping with uncertainty in ecological advice: lessons from fisheries. *Trends in Ecology and Evolution* 18: 617–22.

Hassell, M.P., J. Latto, and R.M. May. 1989. Seeing the wood for the trees: detecting density-dependence from existing life-table studies. *Journal of Animal Ecology* 58: 883–92.

Hastings, A. 1997. *Population Biology. Concepts and Models.* Springer-Verlag, New York.

Heide-Jorgensen, M.P. and T. Harkonen. 1992. Epizootiology of the seal disease in the eastern North Sea. *Journal of Applied Ecology* 29: 99–107.

Heppell, S.S., H. Caswell, and L.B. Crowder. 2000. Life histories and elasticity patterns: perturbation analysis for species with minimal demographic data. *Ecology* 81: 654–65.

Higgins, S.I., W.J. Bond, and W.S.W. Trollope. 2000. Fire, resprouting and variability: a recipe for grass-tree coexistence in savanna. *Journal of Ecology* 88: 213–29.

Hilborn, R. and M. Mangel. 1997. *The Ecological Detective. Confronting Models with Data.* Princeton University Press, Princeton, NJ.

Hilborn, R. and C.J. Walters. 1992. *Quantitative Fisheries Stock Assessment.* Chapman & Hall, London. (Comprehensive review of procedures and policies used for managing fish stocks.)

Hilborn, R., C.J. Walters, and D. Ludwig. 1995. Sustainable exploitation of renewable resources. *Annual Review of Ecology and Systematics* 26: 45–67.

Hitchins, P.M. and J.M. Anderson. 1983. Reproductive characteristics and management of the black rhinoceros in the Hluhluwe/Corridor/Umfolozi Game Reserve Complex. *South African Journal of Wildlife Research* 13: 78–85.

Hobbs, N.T. and R. Hilborn. 2006. Alternatives to statistical hypothesis testing in ecology: a guide to self-teaching. *Ecological Applications* 16: 5–19.

Holling, C.S. 1965. The functional response of predators to prey density and its role in mimicry and population regulation. *Memoirs of the Entomological Society of Canada* 45: 1–60.

Holling, C.S. (ed.). 1978. *Adaptive Environmental Assessment and Management.* John Wiley, Chichester.

Hudson, P.J., A.P. Dobson, and D. Newborn. 1998. Preventions of population cycles by parasite removal. *Science* 282: 2256–8.

Hudson, P.J., A. Rizzoli, B.T. Grenfell, H. Heesterbroek, and A.P. Dobson (eds). 2002. *The Ecology of Wildlife Diseases.* Oxford University Press, Oxford.

Illius, A.W. and I.J. Gordon. 1998. Scaling up from functional response to numerical response in vertebrate herbivores. In H. Olff, V.K. Brown, and R.T. Drent (eds), *Herbivores, Plants and Predators*, pp. 397–427. Blackwell Science, Oxford.

Jeltsch, F., S.J. Milton, W.R.J. Dean, and N. van Rooyen. 1997a. Analysing shrub encroachment in the southern Kalahari: a grid-based approach. *Journal of Applied Ecology* 34: 1497–509.

Jeltsch, F., S.J. Milton, W.R.J. Dean, and N. van Rooyen. 1997b. Simulated pattern formation around artificial waterholes in the semi-arid Kalahari. *Journal of Vegetation Science* 8: 177–88.

Jenkins, S.R., B.D. Perry, and W.G. Winkler. 1988. Ecology and epidemiology of raccoon rabies. *Reviews of Infectious Diseases* 10: S620–36.

Johnson, J.B. and K.S. Omland. 2004. Model selection in ecology and evolution. *Trends in Ecology and Evolution* 19: 101–8.

Karels, T.J. and R. Boonstra. 2000. Concurrent density dependence and independence in populations of arctic ground squirrels. *Nature* 408: 460–1.

Korotkov, V.N., D.O. Logofet, and M. Loreau. 2001. Succession in mixed boreal forest in Russia: Markov models and non-Markov effects. *Ecological Modelling* 142: 25–38.

Krebs, C.J., S. Boutin, R. Boonstra, et al. 1995. Impact of food and predation on the snowshoe hare cycle. *Science* 269: 1112–4.

Krebs, C.R., M.S. Gaines, B.L. Keller, J.H. Myers, and R.H. Tamarin. 1973. Population cycles in small rodents. *Science* 179: 35–41.

Krebs, J.R. 1971. Territory and breeding density in the great tit. *Ecology* 52: 2–22.

Krebs, J.R. and N.B. Davies. 1997. *Behavioural Ecology: An Evolutionary Approach.* Blackwell, Oxford.

Lacy, R.C. and T. Kreeger. 1992. *Vortex Users Manual. A Stochastic Simulation of the Extinction Process.* Chicago Zoological Society, Chicago, IL.

Lamberson, R.H., B.R. Noon, C. Voss, and R.J. McKelvey. 1994. Reserve design for territorial species: the effects of patch size and spacing on the viability of the northern spotted owl. *Conservation Biology* 8: 185–95.

Lande, R., S. Engen, and B.-E. Saether. 2003. *Stochastic Population Dynamics in Ecology and Conservation.* Oxford University Press, Oxford, UK. (Outlines the theoretical framework for sustaining harvests in variable environments.)

Lande, R., B.-E. Saether, and S. Engen. 1997. Threshold harvesting for sustainability of fluctuating resources. *Ecology* 78: 1341–50.

Larter, N.C., A.R.E. Sinclair, T. Ellsworth, J. Nishi, and C.C. Gates. 2000. Dynamics of reintroduction of an indigenous large ungulate: the wood bison of northern Canada. *Animal Conservation* 4: 299–309.

Lawes, M.J., S.E. Piper, and P. Mealin. 2000. Patch occupancy and potential metapopulation dynamics of three forest mammals in fragmented afromontane forest in South Africa. *Conservation Biology* 14: 1088–98.

Laws, R.M. 1969. The Tsavo Research Project. *Journal of Reproduction and Fertility, Suppl.* 6: 495–531.

Laws, R.M., I.S.C. Parker, and R.C.B. Johnstone. 1975. *Elephants and Their Habitats. The Ecology of Elephants in North Bunyoro, Uganda.* Oxford University Press, Oxford.

Levins, R. 1969. Some demographic and genetic consequences of environmental heterogeneity for biological control. *Bulletin of the Entomological Society of America* 15: 237–40.

Lewis, M. and P. Kareiva. 1993. Allee dynamics and the spread of invading organisms. *Theoretical Population Biology* 43: 141–58.

Lidicker, W.Z. 1962. Emigration as a possible mechanism permitting the regulation of population density below carrying capacity. *American Naturalist* 96: 29–33.

Lidicker, W.Z. 1975. The role of dispersal in the population ecology of small mammals. In F.B. Golley, K. Petrusewicz, and L. Ryszkowski (eds), *Small Mammals: Their Productivity and Population Dynamics*, pp. 103–28. Cambridge University Press, Cambridge.

Lierman, M. and R. Hilborn. 1997. Depensation in fish stocks: a hierarchic Bayesian meta-analysis. *Canadian Journal of Fisheries and Aquatic Sciences* 54: 1976–84.

Lima, S.L. and P.A. Zollner. 1996. Towards a behavioural ecology of ecological landscapes. *Trends in Ecology and Evolution* 11: 131–5.

Lindenmayer, D.B., M.A. Burgman, H.R. Akcakaya, R.C. Lacy, and H.P. Possingham. 1995. A review of the generic computer programs ALEX, RAMAS/space, and VORTEX for modelling the viability of wildlife populations. *Ecological Modelling* 82: 161–74.

Lindeque, P.M. and P.C.B. Turnbull. 1994. Ecology and epidemiology of anthrax in the Etosha National Park, Namibia. *Onderstepoort Journal of Veterinary Research* 61: 71–83.

Livezey, B.C. 1993. An ecomorphological review of the dodo and the solitaire, flightless Columbiformes of the Mascarene Islands. *Journal of Zoology, London* 230: 247–92.

Lloyd-Smith, J.O., P.C. Cross, C.J. Biggs, et al. 2005b. Should we expect population thresholds for wildlife disease? *Trends in Ecology and Evolution* 20: 511–9.

Lloyd-Smith, J.O., S.J. Schreiber, P.E. Kopp, and W.M. Getz. 2005a. Superspreading and the effect of individual variation on disease emergence. *Nature* 438: 355–9.

Lotka, A.J. 1925. *Elements of Mathematical Biology.* Dover, New York.

Ludwig, D. 1998. Management of stocks that may collapse. *Oikos* 83: 397–402.

Ludwig, D., R. Hilborn, and S.C. Walters. 1993. Uncertainty, resource exploitation and conservation: lessons from history. *Science* 260: 17, 36.

MacArthur, R.H. and E.O. Wilson. 1963. An equilibrium theory of insular biogeography. *Evolution* 17: 373–87.

MacArthur, R.H. and E.O. Wilson. 1967. *The Theory of Island Biogeography.* Princeton University Press, Princeton, NJ.

Mace, G.M. and R. Lande. 1991. Assessing extinction threats: towards a reevaluation of IUCN threatened species categories. *Conservation Biology* 5: 1148–57.

Maddox, J. 1993. Bringing the extinct dodo back to life. *Nature* 365: 291.

Manly, B.F.J. 1990. *State-structured Populations: Sampling, Analysis and Simulation.* Chapman and Hall, New York.

Manly, B.F.J., L.L. McDonald, D.L. Thomas, T.L. McDonald, and W.P. Erickson. 2002. *Resource Selection by Animals: Statistical Design and Analysis for Field Studies.* Chapman & Hall, London.

May, R.M. 1981. The role of theory in ecology. *American Zoologist* 21: 903–10.

Mayaka, T.B., J.D. Stigter, I.M.A. Heitkonig, and H.H.T. Prins. 2004. A population dynamics model for the management of Buffon's kob in the Benoue National Park Complex, Cameroon. *Ecological Modelling* 176: 135 53.

McCallum, H. 2000. *Population Parameters. Estimation for Ecological Models.* Blackwell, Oxford.

McCallum, H. and A. Dobson. 1995. Detecting disease and parasite threats to endangered species and ecosystems. *Trends in Ecology and Evolution* 10: 190–4.

McCallum, H., N. Barlow, and J. Hone. 2001. How should pathogen transmission be modelled? *Trends in Ecology and Evolution* 16: 295–300.

McCarthy, M.A., H.P. Possingham, J.R. Day, and A.J. Tyne. 2001. Testing the accuracy of population viability analysis. *Conservation Biology* 15: 1030–8.

McCullough, D.R. 1992. Concepts of large herbivore population dynamics. In D.R. McCullough and R.H. Barrett (eds), *Wildlife 2001: Populations*, pp. 967–84. Elsevier, London, UK.

Mduma, S.A.R., A.R.E. Sinclair, and R. Hilborn. 1999. Food regulates the Serengeti wildebeest: a 40-year record. *Journal of Animal Ecology* 68: 1101–22.

McLoughlin, C.A. and N. Owen-Smith. 2003. Viability of a diminishing roan antelope population: predation is the threat. *Animal Conservation* 6: 231–6.

McShea, W.J., H.B. Underwood, and J.H. Rappole (eds). 1997. *The Science of Overabundance. Deer Ecology and Population Management.* Smithsonian Institute Press, Washington, DC.

Mduma, S., R. Hilborn, and A.R.E. Sinclair. 1998. Limits to exploitation of Serengeti wildebeest and implications for its management. In D.M. Newbery, H.H.T. Prins, and N.D. Brown (eds), *Dynamics of Tropical Communities*, pp. 243–65. Blackwell, Oxford.

Mills, M.G.L., H.C. Biggs, and I.J. Whyte. 1995. The relationship between rainfall, lion predation and population trends in African herbivores. *Wildlife Research* 22: 75–88.

Milner-Gulland, E.J. and R. Mace. 1998. *Conservation of Biological Resources*. Blackwell, Oxford.

Moore, A.D. 1990. The semi-Markov process: a useful tool in the analysis of vegetation dynamics for management. *Journal of Environmental Management* 30: 111–30.

Morris, W.F. and D.F. Doak. 2002. *Quantitative Conservation Biology*. Sinauer Associates, Sunderland, MA.

Myers, R.A., N.J. Barrowman, J.A. Hutchings, and A.A. Rosenberg. 1995. Population dynamics of exploited fish stocks at low population levels. *Science* 269: 1106–8.

Newman, E.I. 1993. *Applied Ecology*. Blackwell Science, Oxford.

Nichols, J.D., F.A. Johnson, and B.K. Williams. 1995. Managing North American waterfowl in the face of uncertainty. *Annual Review of Ecology and Systematics* 26: 177–99.

Noon, B.R. and K.S. McKelvey. 1996. A common framework for conservation planning: linking individual and metapopulation models. In D. McCullough (ed.), *Metapopulations and Wildlife Conservation*, pp. 139–66. Island Press, Washington, DC.

Noon, B.R. and J.R. Sauer. 1992. Population models for passerine birds: structure, parameterization and analysis. In D.R. McCullough and R.H. Barrett (eds), *Wildlife 2001: populations*, pp. 441–64. Elsevier, London.

Norris, K. 2004. Managing threatened species: the ecological toolbox, evolutionary theory and the declining-population paradigm. *Journal of Applied Ecology* 41: 413–26.

Norris, K. and R. Stillman. 2002. Predicting the impact of environmental change. In K. Norris and D.J. Pain (eds), *Conserving Bird Biodiversity*, pp. 202–23. Cambridge University Press, Cambridge.

Norton, P.M. 1994. Simple spreadsheet models to study population dynamics, as illustrated by a mountain reedbuck model. *South African Journal of Wildlife Research* 24: 73–81.

Ogutu, J. and N. Owen-Smith. 2003. ENSO, rainfall and temperature influences on extreme population declines among African savanna ungulates. *Ecology Letters* 6: 412–9.

Osho, J.S.A. 1991. Matrix model for tree population projection in a tropical rain forest of south-western Nigeria. *Ecological Modelling* 59: 247–55.

Osho, J.S.A. 1995. Optimal sustainable harvest models for a Nigerian tropical rain forest. *Journal of Environmental Management* 45: 101–8.

Ottersen, G., B. Planque, A. Belgrano, E. Post, P.C. Reid, and N.C. Stenseth. 2001. Ecological effects of the North Atlantic Oscillation. *Oecologia* 128: 1–14.

Owen-Smith, N. 1981. The white rhinoceros overpopulation problem, and a proposed solution. In P.A. Jewell, S. Holt, and D. Hart (eds), *Problems in Management of Locally Abundant Wild Mammals*, pp. 129–50. Academic Press, New York.

Owen-Smith, N. 1983. Dispersal and the dynamics of large herbivore populations in enclosed areas. In R.N. Owen-Smith (ed.), *Management of Large Mammals in African Conservation Areas*, pp. 127–43. Haum, Pretoria.

Owen-Smith, N. 1988. *Megaherbivores. The Influence of Very Large Body Size on Ecology*. Cambridge University Press, Cambridge.

Owen-Smith, N. 1990. Demography of a large herbivore, the greater kudu, in relation to rainfall. *Journal of Animal Ecology* 59: 893–913.

Owen-Smith, N. 1993a. Evaluating optimal diet models for an African browsing ruminant, the kudu: how constraining are the assumed constraints? *Evolutionary Ecology* 7: 499–524.

Owen-Smith, N. 1993b. Comparative mortality rates of male and female kudus: the costs of sexual size dimorphism. *Journal of Animal Ecology* 62: 428–40.

Owen-Smith, N. 1994. Foraging responses of kudus to seasonal changes in food resources: elasticity in constraints. *Ecology* 75: 1050–62.

Owen-Smith, N. 1998. How high ambient temperature affects the daily activity and foraging time of a subtropical ungulate, the greater kudu. *Journal of Zoology* 246: 183–92.

Owen-Smith N. 2000. Modeling the population dynamics of a subtropical ungulate in a variable environment: rain, cold and predators. *Natural Resource Modeling* 13: 57–87.

Owen-Smith, N. 2002a. Credible models for herbivore – vegetation systems: towards an ecology of equations. *South African Journal of Science* 98: 445–9.

Owen-Smith, N. 2002b. *Adaptive Herbivore Ecology. From Resources to Populations in Variable Environments.* Cambridge University Press, Cambridge.

Owen-Smith, N. 2003. Foraging behavior, habitat suitability, and translocation success, with special reference to large mammalian herbivores. In M. Festa-Bianchet and M. Apollonio (eds), *Animal Behavior and Wildlife Management*, pp. 93–109. Island Press, Washington, DC.

Owen-Smith, N. 2005. Incorporating fundamental laws of biology and physics into population ecology: the metaphysiological approach. *Oikos* 111: 611–15.

Owen-Smith, N. 2006. Demographic determination of the shape of density dependence for three African ungulate populations. *Ecological Monographs* 76: 73–92.

Owen-Smith, N. and S.M. Cooper. 1987. Palatability of woody plants to browsing ungulates in a South African savanna. *Ecology* 68: 319–31.

Owen-Smith, N. and S.M. Cooper. 1989. Nutritional ecology of a browsing ruminant, the kudu, through the seasonal cycle. *Journal of Zoology* 219: 29–43.

Owen-Smith, N., G.I.H. Kerley, B. Page, R. Slotow, and R.J. van Aarde (in press). A scientific perspective on the management of elephants in the Kruger National Park and elsewhere. *South African Journal of Science*.

Owen-Smith, N. and M.G.L. Mills. 2006. Manifold interactive influences on the population dynamics of a multi-species ungulate assemblage. *Ecological Monographs* 76: 73–92.

Owen-Smith, N., D.R. Mason, and J.O. Ogutu. 2005. Correlates of survival rates for ten African ungulate populations: density, rainfall and predation. *Journal of Animal Ecology* 74: 774–88.

Pain, D.J. and P.F. Donald. 2002. Outside the reserve: pandemic threats to bird biodiversity. In K. Norris and D.J. Pain (eds), *Conserving Bird Biodiversity*, pp. 157–79. Cambridge University Press, Cambridge.

Parsons, M., C.A. McLoughlin, K.A. Kotschy, K.H. Rogers, and M.W. Roundtree. 2005. The effects of extreme floods on the biophysical heterogeneity of river landscapes. *Frontiers in Ecology and the Environment* 3: 487–94.

Persson, L. and A.M. De Roos. 2003. Adaptive habitat use in size-structured populations: linking individual behavior to population processes. *Ecology* 84: 1129–39.

Peters, C.M. 2001. Lessons from the plant kingdom for conservation of exploited species. In J.D. Reynolds, G.M. Mace, K.H. Redford, and J.G. Robinson (eds), *Conservation of Exploited Species,* pp. 242–56. Cambridge University Press, Cambridge.

Pettifor, R.A., R.W.G. Caldow, J.M. Rowcliffe, et al. 2000a. Spatially explicit, individual-based, behavioural models of the annual cycle of two migratory goose populations. *Journal of Applied Ecology* 37 (suppl. 1): 103–35.

Pettifor, R.A., K.J. Norris, and J.M. Rowcliffe. 2000b. Incorporating behaviour in predictive models for conservation. In L.M. Gosling and W.J. Sutherland (eds), *Behaviour and Conservation,* pp. 198–220. Cambridge University Press, Cambridge.

Plowright, W. 1982. The effects of rinderpest and rinderpest control on wildlife in Africa. *Symposium of the Zoological Society of London* 50: 1–28.

Prout, T. and F. McChesney. 1985. Competition among immatures affects their adult fertility: population dynamics. *American Naturalist* 126: 521–58.

Punt, A.E. and A.D.M. Smith. 2001. The gospel of maximum sustainable yield in fisheries management: birth, crucifixion and reincarnation. In J.D. Reynolds, G.M. Mace, and J.G. Robinson (eds), *Conservation of Exploited Species,* pp. 41–66. Cambridge University Press, Cambridge.

Randolph, S.E., C. Chemini, C. Furnanello, et al. 2001. The ecology of tick-born infections in wildlife reservoirs. In P.J. Hudson, A. Rizzoli, B.T. Grenfell, H. Heesterbroek, and A.P. Dobson (eds), *The Ecology of Wildlife Diseases,* pp. 119–38. Oxford University Press. Oxford.

Ralls, K., K. Brugger, and J. Ballou. 1979. Inbreeding and juvenile mortality in small populations of ungulates. *Science* 206: 1101–3.

Reed, J.M., L.S. Mills, J.B. Dunning, et al. 2002. Emerging issues in population viability analysis. *Conservation Biology* 16: 7–19.

Reynolds, J.D. and S. Jennings. 2000. The role of animal behaviour in marine conservation. In L.M. Gosling and W.J. Sutherland (eds), *Behaviour and Conservation,* pp. 238–60. Cambridge University Press, Cambridge.

Reynolds, J.D., G.M. Mace, K.H. Redford, and J.G. Robinson. 2001. *Conservation of Exploited Species.* Cambridge University Press, Cambridge.

Riney, T. 1964. The impact of introductions of large herbivores on the tropical environment. IUCN public., new series no. 4, pp. 261–73.

Roberts, M.G. 1996. The dynamics of bovine tuberculosis in possum populations, and its eradication or control by culling or vaccination. *Journal of Animal Ecology* 65: 451–64.

Roberts, C.M. 1997. Ecological advice for the global fisheries crisis. *Trends in Ecology and Evolution* 12: 35–8.

Roelke-Parker, M.E., L. Muson, C. Pacher, et al. 1996. A canine distemper virus epidemic in Serengeti lions. *Nature* 381: 172.

Rogers, K.H. and J. O'Keefe. 2003. River heterogeneity: ecosystem structure, function, and management. In J.T. du Toit, K.H. Rogers, and H.C. Biggs (eds), *The Kruger Experience,* pp. 189–218. Island Press, Washington, DC.

Rosenzweig, M.L. 1971. Paradox of enrichment: destabilization of exploitation systems. *Science* 171: 385–7.

Rosenzweig, M.L. and Z. Abramsky. 1997. Two gerbils of the Negev: a long-term investigation of optimal habitat selection and its consequences. *Evolutionary Ecology* 11: 733–56.

Rosenzweig, M.L. and R.H. MacArthur. 1963. Graphical representation and stability conditions of predator–prey interaction. *American Naturalist* 97: 209–23.

Roughgarden, J. 1998. *Primer of Ecological Theory*. Prentice-Hall, NJ.

Rountree, M.W., K.H. Rogers, and G.L. Heritage. 2000. Landscape state change in the semi-arid Sabie River, Kruger National Park, in response to flood and drought. *South African Geographical Journal* 82: 173–81.

Rutherford, M.C. 1980. Annual plant production – precipitation relations in arid and semi-arid regions. *South African Journal of Science* 76: 53–6.

Saether, B.-E., S. Engen, A.P. Moller, et al. 2004. Life-history variation predicts the effects of demographic stochasticity on avian population dynamics. *American Naturalist* 164: 793–802.

Sauer, J.R. and N.A. Slade. 1987. Size-based demography of vertebrates. *Annual Review of Ecological Systems* 18: 71–90.

Shaffer, M.L. 1981. Minimum population sizes for species conservation. *BioScience* 31: 131–4.

Sillero-Zubiri, C., A.A. King, and D.W. MacDonald. 1996. Rabies and mortality in Ethipian wolves. *Journal of Wildlife Diseases* 32: 80–6.

Silvertown, J.W. and J.L. Doust. 1993. *Introduction to Plant Population Biology*. Blackwell, Oxford.

Sinclair, A.R.E., J.M. Fryxell, and G. Caughley. 2005. *Wildlife Ecology, Conservation and Management*, 2nd edn. Blackwell, Oxford.

Sjogren-Gulve, P. and T. Eberhard (eds). 2000. *The Use of Population Viability Analysis in Conservation Planning*. Wallin & Dolholm, Lund, Sweden.

Smuts, G.L. 1978. Interrelations between predators, prey, and their environment. *BioScience* 28: 316–20.

Starfield, A.M. 1990. Qualitative, rule-based modeling. *BioScience* 40: 601–4.

Starfield, A.M. 1997. A pragmatic approach to modeling for wildlife management. *Journal of Wildlife Management* 61: 261–70.

Starfield, A.M. and A.L. Bleloch. 1991. *Building Models for Conservation and Wildlife Management*, 2nd edn. Burgess, Edina, MN.

Starfield, A.M., D.H.M. Cumming, R.D. Taylor, and M.S. Quadling. 1993. A frame-based paradigm for dynamic ecosystem models. *AI Applications* 7: 1–13.

Stephens, P.A. and W.J. Sutherland. 1999. Consequences of the Allee effect for behaviour, ecology and conservation. *Trends in Ecology and Evolution* 14: 401–5.

Stillman, R.A., J.D. Goss-Custard, A.D. West, et al. 2001. Predicting shorebird mortality and population size under different regimes of shellfishery management. *Journal of Applied Ecology* 38: 857–68.

Strickland, H.E. and A.G. Melville. 1848. *The Dodo and Its Kindred; or the History, Affinities and Osteology of the Dodo, Solitaire, and Other Extinct Birds of the Islands Mauritius, Rodriguez, and Bourbon*. Reeve, Behan and Reeve, London.

Sutherland, W.J. 1996. *From Individual Behaviour to Population Ecology*. Oxford University Press, Oxford.

Sutherland, W.J. and G.A. Allport. 1994. A spatial depletion model of the interaction between bean geese and wigeon with the consequences for habitat management. *Journal of Animal Ecology* 63: 51–9.

Sutherland, W.J. and K. Norris. 2002. Behavioural models of population growth rates: implications for conservation and prediction. *Philosophical Transactions of the Royal Society of London, Series B* 357: 1273–84.

Swinton, J., M.E.J. Woolhouse, M.E. Begon, et al. 2002. Microparasite transmission and persistence. In P.J. Hudson, A. Rizzoli, B.T. Grenfell, H. Heesterbroek, and A.P. Dobson (eds), *The Ecology of Wildlife Diseases*, pp. 83–101. Oxford University Press, Oxford.

Swinton, J., J. Harwood, B.T. Grenfell, and J. Harwood. 1998. Persistence threshold for phocine distemper virus infection in harbour seal metapopulations. *Journal of Animal Ecology* 67: 54–68.

Taper, M.L. and P.J.P. Gogan. 2002. The northern Yellowstone elk herd: density dependence and climatic conditions. *Journal of Wildlife Management* 66: 106–22.

Thomas, C.D. and I. Hanski. 1997. Butterfly metapopulations. In I. Hanski and M. Gilpin (eds), *Metapopulation Biology*, pp. 359–86. Academic Press, San Diego, CA.

Thorne, E.T. and E.S. Williams. 1988. Disease and endangered species: the black-footed ferret as a recent example. *Conservation Biology* 2: 66–74.

Turchin, P.L. 2003. *Complex Population Dynamics. A Theoretical/Empirical Synthesis.* Princeton University Press, Princeton, NJ.

Usher, M.M. 1981. Modelling ecological succession, with particular reference to Markovian models. *Vegetatio* 46: 11–8.

Van Aarde, R.J., T.P. Jackson, and S.M. Ferreira (in press). Conservation science and elephant management in southern Africa. *South African Journal of Science.*

Vandermeer, J.H. and D.E. Goldberg. 2003. *Population Ecology. First Principles.* Princeton University Press, Princeton, NJ.

Volterra, V. 1926. Fluctuations in the abundance of species considered mathematically. *Nature* 118: 558–60.

Walker, B.H., R.H. Emslie, N. Owen-Smith, and R.J. Scholes. 1987. To cull or not to cull: lessons from a southern African drought. *Journal of Applied Ecology* 24: 381–402.

Walker, B.H., D. Ludwig, C.S. Holling, and R.S. Peterman. 1981. Stability of semi-arid savanna grazing systems. *Journal of Ecology* 69: 473–98.

Walters, C. 1986. *Adaptive Management of Renewable Resources.* Macmillan. (A thorough valuation of the principles involved in managing bio-resources adaptively, with a focus on fisheries.)

Walters, C.J., R. Hilborn, and R.M. Peterson. 1975. Computer simulation of barren-ground caribou dynamics. *Ecological Modelling* 1: 303–15.

Weber, G.E., F. Jeltsch, N. van Rooyen, and S.J. Milton. 1998. Simulated long-term vegetation response to grazing heterogeneity in semi-arid rangelands. *Journal of Applied Ecology* 35: 687–99.

Westoby, M., B.H. Walker, and I. Noy-Meir. 1989. Opportunistic management for rangelands not at equilibrium. *Journal of Range Management* 42: 266–74.

White, G.C. 2000. Population viability analysis: data requirements and essential analysis. In L. Boitani and T.K. Fuller (eds), *Research Techniques in Animal Ecology*, pp. 288–331. Columbia University Press, New York.

Whyte, I.J., R. van Aarde, and S.L. Pimm. 1998. Managing the elephants of the Kruger National Park. *Animal Conservation* 1: 77–83.

Wiegand, K., F. Jeltsch, and D. Ward. 2004. Minimum recruitment frequency in plants with episodic recruitment. *Oecologia* 141: 363–72.

Wiegand, T., S.J. Milton, and C. Wissel. 1995. A simulation model for a shrub ecosystem in the semiarid Karoo, South Africa. *Ecology* 76: 2205–21.

Williams, B.G. and C.M. Campbell. 1996. A model of HIV transmission on South African mines: implications for control. *South African Journal of Epidemiology and Infection* 11: 51–5.

Williams, B.G., E. Gouws, M. Colvin, F. Sitas, G. Ramjee, and A.S.A. Karim. 2000a. Patterns of infection: using age prevalence data to understand the epidemic of HIV in South Africa. *South African Journal of Science* 96: 305–12.

Williams, B.G., E. Gouws, and A.S.A. Karim. 2000b. Where are we now? Where are we going? The demographic impact of HIV/AIDS in South Africa. *South African Journal of Science* 96: 297–300.

Williams, B.G., C. MacPhail, C. Campbell, et al. 2000c. The Carletonville–Mothusimpilo Project: limiting transmission of HIV through community-based interventions. *South African Journal of Science* 96: 351–9.

Williams, B.K., F.A. Johnson, and K. Wilkins. 1996. Uncertainty and the adaptive management of waterfowl harvests. *Journal of Wildlife Management* 60: 223–32.

Williams, B.K., J.D. Nichols, and M.J. Conroy. 2002. *Analysis and Management of Animal Populations.* Academic Press, San Diego, CA.

Wilson, K., O.N. Bjornstad, A.P. Dobson, et al. 2002. Heterogeneities in macroparasite infections: parasites and processes. In P.J. Hudson, A. Rizzoli, B.T. Grenfell, H. Heesterbroek, and A.P. Dobson (eds), *The Ecology of Wildlife Diseases*, pp. 6–44. Oxford University Press, Oxford.

Woiwod, I.P. and I. Hanski. 1992. Patterns of density dependence in moths and aphids. *Journal of Animal Ecology* 61: 619–29.

Index